OHIO ART
THE WORLD OF TOYS

LISA KERR

IN COLLABORATION

WITH JIM GILCHER

WITH PHOTOGRAPHY BY ADAM GRIFFITH

4880 Lower Valley Rd. Atglen, PA 19310 USA

This book is dedicated to my husband, Bruce, and my children, Jacob and Ariel, who keep my heart young.

Designed by Bonnie M. Hensley
Typeset in Humanist 521 Ultrabold/Geometric 415Lt Bt

ISBN: 0-7643-0512-3
Printed in Hong Kong
1 2 3 4

The information in this book has been researched and compiled from reliable sources. Efforts have been made to avoid errors and/or questionable information. However, errors are possible in a work of this scope. Neither the publisher nor the author or collaborator will be held responsible for losses that may occur in either the purchase, sale, or any other transactions that may occur based on the information contained in this book.

Library of Congress Cataloging-in-Publication Data

Kerr, Lisa.
    The Ohio art book / Lisa Kerr ; in collaboration with Jim Gilcher ; with photography by Adam Griffith.
        p. cm.
    Includes index.
    ISBN 0-7643-0512-3
    1. Ohio Art Company--Catalogs. 2. Toys--Collectors and collecting--United States--Catalogs. 3. Toys--United States--History--20th century--Catalogs. I. Gilcher, Jim. II. Griffith, Adam. III. Title.
    NK9509.65.U640385   1998   97-48751
    688.7'2'922--dc21   CIP

Published by Schiffer Publishing Ltd.
4880 Lower Valley Road Atglen, PA 19310
Phone: (610) 593-1777; Fax: (610) 593-2002
e-mail: schifferbk@aol.com Please write for a free catalog.
This book may be purchased from the publisher.
Please include $3.95 for shipping.

Please try your bookstore first.
We are interested in hearing from authors
with book ideas on related subjects.

TABLE OF CONTENTS

ACKNOWLEDGMENTS

Many thanks to all those who contributed time and energy to creating this book. To Doug and Pat Wengel for making their toys available to me and for their high spirit and encouragement. To Bob Bernabe, for bringing toys to the photo shoot in New Jersey. To William C. Killgallon, President and CEO, Ohio Art Company, for his time and hospitality. To his executive secretary, Betty Cole for her patience and efficiency. To John Walters of the Ohio Art Company for his enthusiastic participation in the making of this book. To Sharon Lazane, Susan L. Phillips, Claudette Job, and Shirley Cox for sharing their information and/or collections. To John Griffith for his tireless grammatical corrections.

INTRODUCTION

On the last day of October 1996, I visited the Ohio Art toy company in Bryan, Ohio. I was filled with nervous anticipation; I had a certain old-fashioned image of the company that I was afraid might be ruined by an actual visit. Happily, I was not disappointed.

It was night when we first arrived and the streets were clean, well-lit, and quiet. Winter was already raising its gray Midwestern flag. A chilly wind unfurled down Main Street, folded around corners, and sent leaves flying down the starched streets.

As we drove into the town center a fantastic structure loomed up before us. It was the Williams County Courthouse, built in 1890. This towered and turreted giant reigns over this straightforward Midwestern town with an exotic, Moorish flair. Surrounding the courthouse is a public park with a bandstand and large graceful trees. The park is still used for religious, political, and social gatherings. Even on this cold October day, it was easy to picture a summer night, decades ago, when Bryan citizens strolled through the park, eating ice cream and listening to the Bryan City Band play, just as it has played every summer night since 1902.

The next morning, bright and early, we appeared at the old brick building that has housed Ohio Art for over sixty-five years. After a brief visit, Jim Gilcher took us to meet Mr. Killgallon, the president and CEO of Ohio Art. He was a gracious and generous host. In fact, everyone was incredibly helpful and willing to offer any information they had. From speaking with local citizens in cafes and shops, I had already began to realize what an institution Ohio Art was in Bryan. So many people seemed to either have a relative or know someone who had once worked for the company. It was not too long ago that Ohio Art was the place for high-school and college students to earn money during the summer break. This has since changed, and as I spoke with Mr. Killgallon, I began to realize that time could not stand still for Ohio Art. The company had to stay competitive to stay in business, which meant incorporating many of the manufacturing methods used by other American companies such as shifting production overseas. I was glad to see, however, that some toys were still produced in-house and was pleased to be given an opportunity to observe an assembly line for "Etch A Sketch" and some of the lithography processes. Despite the changes Ohio Art has made, it remains a major contributor to the economy of Bryan, Ohio.

Although there were several American companies producing tin-litho toys from the 1920s through the 1980s, the quality of Ohio Art toys and the beauty of the lithography outshines all the others. Ohio Art was a stickler for quality. If a piece was flawed, it did not pass quality control. The company employed a wider range of colors than found on most other lithographed toys and only the best metal was used. This accounts, but only in part, for the beauty of much of the lithography. Artistry, good taste, and hard work account for the rest.

This book attempts to cover a wide range of toys produced by Ohio Art. It is by no means all inclusive. Ohio Art produced many more toys than are referenced on these pages, but we have attempted to include toys that we believe are especially desirable to collectors as well as toys that represent a change or new direction for the company. Because the toys of the 1920s through the 1950s are the most collectible, they are most heavily represented. However, there are many highly collectible tea sets, sand pails, tops, drums, etc. that are not pictured in this book. Usually this is because we were unable to obtain a photograph that was of publishing quality. Sometimes, however, we were surprised at the last minute by a toy we did not even know existed! It is this potential for surprise that makes collecting Ohio Art toys so much fun. Hopefully, with the publication of this book, heretofore unknown toys will surface and assume their place in the Ohio Art legacy.

Williams County Courthouse

CHAPTER I: PICTURE THIS

THE BEGINNING

On February 1, 1876, Henry S. Winzeler, one of six children, was born to Michael and Magdalene Winzeler of Archbold, Ohio. Like his siblings, Henry possessed great mechanical ability and an innovative spirit. He worked as a teacher and store-keeper. After attending dental school, Henry Winzeler opened a dentist office in the Murbach Building in Archbold.

But Henry required more stimulation than the practice of dentistry could provide. He was restless to pursue new ideas. On one of his frequent visits to the local clothing store owned by his aunt Christina Lauber, Henry, who had a special love for art, was drawn to the oval mirror mounted on a frame which stood on the counter. The mirror was used by gentlemen selecting hats. Admiring the shape of the mirror, Henry was struck by the idea of what an attractive shape the oval would make as a frame for beautiful pictures. The frame could be covered with glass and supplied with a fastener for hanging on the wall. Henry Winzeler translated his idea into actions. On October 6, 1908, he formed a com-

Henry S. Winzeler, founder of the Ohio Art Company

pany to manufacture metal picture frames. The frames were stamped out in Toledo and shipped to Archbold in barrels where they were spray painted and assembled. From there, they were packaged in cardboard, placed in wooden crates, many of which were handmade by Henry's father, picked up by a horse-drawn dray, and taken to the train station for shipping. The picture frame manufacturing concern was aptly name the OHIO ART COMPANY.

On October 8, 1908, the local newspaper, *The Archbold Buckeye*, reported that Ohio Art employed eight workers. The workforce quickly multiplied to meet growing demands for the frames, and by 1910 the *Buckeye* printed a photo that showed that the Ohio Art workforce had increased threefold. However, this was just the beginning. The picture frames really caught on and were marketed through retailers including Woolworth's, Kresge, Sears, and Butler Brothers. The Ohio Art Company kept on growing until finally, running out of available space in Archbold, the company

The Ohio Art building constructed in Bryan, Ohio in 1912. The building was 13,500 square feet. Railroad siding, capable of handling fourteen large freight cars, ran right by the building.

moved to new quarters in the nearby town of Bryan, Ohio. By 1914, the company employed sixty-five people.

The new building in Bryan covered 13,500 square feet which included office and manufacturing space. To create a homey touch to the factory, trees and flower beds were planted around the building, tended by Ms. Winzeler. In 1915, the company installed the first lithography equipment to produce the wood-grained metal sheets which were formed into the picture frames. Henry Winzeler was very well regarded in the business world and enjoyed a reputation for prompt and careful business practices.

Ohio Art "Old-Timers." Pictured are some of Henry Winzeler's contemporaries who knew him as a neighbor and a colleague.

BEYOND THE FRAME

In 1917, seeking to diversify, Henry Winzeler purchased the C.E. Carter Company located in Erie, Pennsylvania. Carter produced a galvanized toy windmill as well as a climbing monkey on a string for Ferdinand Strauss of New York City. After World War I, Louis Marx approached Henry Winzeler to sell him the Erie factory. The story goes that Henry was nailing tile onto the roof of his new home. Mr. Marx climbed a ladder to the roof to join him and the deal was struck. The Erie plant became one of the foundations for the Marx Toy Company and Henry and Louis Marx remained fast friends.

It was in 1918 that the Ohio Art Company produced its first toy in Bryan, Ohio—a lithographed metal tea set. Henry displayed these first tea sets, adorned with the ABCs around the outer rim of the plates, at the March 9, 1920, New York Toy Fair. In 1923, he introduced a metal sand pail line at the show; the line included metal lithographed pails, shovels, and sand molds. By 1923, the Lithography Department was in production twenty-two hours a day and employed over 200 people. The factory temporarily closed in 1927 to install new machinery to accomodate an increase in tea set production. By then, employment was up to 250.

Throughout the company's growth, Henry's holdings continued to grow and evolve. In 1920, the Bryan plant announced an expansion to accommodate the recently purchased assets of the Holabird Manufacturing Company of Chicago, Illinois. Henry also purchased the Battle Creek Toy Manufacturing Company of Battle Creek, Michigan, in 1917. The business was later moved to Bryan.

In 1930 Ohio Art acquired assets of Mutual Novelty Manufacturing Company, which originated the process for manufacturing Christmas tree icicles. The same year the company acquired the distribution and manufacturing rights of Household Appliance Manufacturing Company, which manufactured clothes dryers, as well as partnership assets of Veelo Manufacturing Company, a manufacturer of stuffed dolls and animals.

THE OHIO ART FAMILIES

Henry Winzeler was a great believer in the "American Dream" (a philosophy which withstands strict scrutiny only if one believes that all people begin life on an even playing field). Henry believed in hard work and raised his children to respect America and the free enterprise system.

Among Mr. Winzeler's private papers were several writings entitled "Poems that have helped me." One in particular stands out as an expression of his entrepreneurial spirit:

We must not hope to be mowers,
And to gather the ripe golden ears
Unless we have first been sowers
And watered the furrows with tears.
-Goethe

In 1913, Mr. Winzeler was blessed with a daughter, Eugenia, and again in 1915 with a son Howard, affectionately known as "Howie." By 1933, Howard was a full-time employee at Ohio Art Toy Company. Howard mastered the intricacies of each department, and his broad spectrum of knowledge was a tremendous asset to the company. In 1939, he became a Director; in 1953, President; and by 1966, he was Chairman of the Board.

Howard Winzeler, son of Henry S. Winzeler.

Mr. Lachlan Malcolm MacDonald became part owner of the Ohio Art Company during the first dark days of the Depression. With Mr. MacDonald's capable guidance and assistance, the company withstood the Depression, continuing to employ hundreds of Bryan residents.

Lachlan Malcolm MacDonald.

William Casley Killgallon joined the Ohio Art Company in 1955 as Vice-president of sales. In 1957, he became a director and in 1966, President. In 1978 he was elected Chairman of the Board. In the tradition of other leaders in the Ohio Art Company, Mr. Killgallon was another highly self-motivated and energetic individual.

William Casley Killgallon.

William Casley Killgallon passed his business acumen and energy along to his son, William Carpenter Killgallon. After serving with Army Intelligence for three years, William Carpenter Killgallon worked for The Bank of New York. He was The National Tour Director for the Nixon family and worked on the Richard M. Nixon presidential campaign staff. He joined Ohio Art in 1968 and was elected President and Chief Executive Officer in 1978, a position he holds to this day.

William Carpenter Killgallon.

A BRIEF TOY HISTORY

The Ohio Art Company played an important role in the lives of Bryan citizens. In 1926, Henry Winzeler offered to donate $25,000 to build a hospital in Bryan if the city matched the sum. By 1927 employment had increased to 250. The company continued to provide a source of employment to residents of Bryan all the way through the 1980s. At times, multiple members and several generations of one family worked at Ohio Art together. In the early 1940s, Ohio Art distinguished itself locally by being the first factory in Bryan to procure 100 percent sign-up for payroll defense bonds. The company produced steel shell casings and rocket parts for the United States military in 1942 and earned the Army-Navy "E" for excellence in war production.

In the early 1950s Ohio Art began using plastic parts, supplied by Champion Molded Plastics, in some of its toy production. In 1953, the company began to manufacture plastic parts in-house. The company added the "Little Chef" line which included miniature cookware sets and electric ranges. Tool boxes, drum sets, mechanical toys, housekeeping items, target sets, lunch boxes, globes, and Roy Rogers items were added to the line. In the late 1950s Ohio Art began distributing the inventory of the Dunwell Company under the tradename "Buckeye Trucks." And then, in 1960, the famous Etch A Sketch appeared on toy shelves across the United States. Today, thirty-seven years later, the only thing that has changed on the Etch A Sketch is the size of the turning knobs!

Up until the end of the 1980s, Ohio Art managed to continue to produce all of its toys here in the United States. The company employed between eight hundred and one thousand people. Recently, in order to remain competitive in the toy business, Ohio Art has switched the majority of its production overseas. However, the full-sized Etch A Sketches and drum sets are still produced in the Bryan Ohio plant. About thirty people are "on line" in the highly automated plant with the capablility of turning out approximately eight thousand Etch A Sketches a day!

Women on the assembly line putting together "Etch A Sketch" toys. Halloween, 1996.

THE LITHOGRAPHY PROCESS

The lithography process begins with an artist's drawing or design. The design is transferred onto a litho printing plate. Paint is rolled onto large flat tin sheets, one color at a time. After the paint is laid down, the sheet is sent through an annealing oven to set the paint. After the color is fixed, the sheet is sent through the process again to add the next color. A new, ultraviolet lithography machine is now available which can lay down six colors, one right after the other. The six colors can be laid down and baked on in sequence in one process. This speeds up the process considerably.

Today, the Ohio Art Company is focusing on developing its litho printing business. The company's goal is to become the dominant lithographer in the United States market, and eventually become a leader in the international market as well. Lithography equipment is a tremendous investment, costing in the neighborhood of six million dollars. Ohio Art uses equipment from the 1930s in conjunction with its modern ultra violet equipment. The steel is manufactured domestically, and this, combined with the availability of skilled artists in the United States, makes a successful international operation a viable goal.

Progressive stages for adding colors in the lithography process.

An example of the heavy-duty equipment used to stamp out and shape toys.

Shaping a shovel:
1st step: sharp-edged flat sheet of metal.
2nd step: shovel is formed, but not curled. The sharp edges are still exposed.
3rd step: all edges curled under so they are safe for children's play.

The Ohio Art Company is well on its way toward achieving its goal of becoming a world leader in color lithography. The company is currently doing all of the advertising lithography for the Triple A Sign company, the series of charming old-fashioned "Coke" trays, and the printing on the Kodak film canisters.

When asked what philosophy best represents the Ohio Art Company today, Mr. Killgallon used the words "a nice honest company to work with." That is something of which the Ohio Art Company deserves to be proud. They are conservative in their approach, maintaining a dress code and encouraging a family spirit, while attempting to keep their operation as domestically based as possible.

PRICING

Of the several factors influencing the pricing of Ohio Art toys, condition is the most important. Condition is everything. I will rarely buy a toy that is rusted although I know several avid collectors who will buy pieces with some rust and hang onto them until they can replace them with pieces in better condition. It really is a matter of personal preference. Of course, a box will elevate the cost of a tin-litho toy considerably. For toys such as sand pails, this is really not an issue because they were generally not packaged in boxes. For a tea set, however, the box is wonderful. It provides a nice display for the set and assures that all of the pieces are from the same lot.

Another influential factor affecting price is geographical location. Older tin-litho toys tend to be more plentiful and more sought after in the midwest and parts of the east coast than on the west coast where toy shows are replete with 1960s, 1970s, and even 1980s toys. Because there are not as many old toys available on the west coast, prices are often erratic, either ridiculously high or far too low.

Tin-litho toys produced under the Walt Disney licenses are, of course, highly sought after. The Ohio Art Walt Disney lithography, especially that produced during the 1930s, is some of the most beautiful there is. Age is another factor that plays an important role. Older toys from the teens, twenties, and thirties will sell for a considerably higher price than toys of the forties and fifties.

For the purpose of this book, the range of prices starts from the condition referred to as "good" at the low-end to excellent at the high end. Good is a toy that has no rust that interferes in any way with the lithography, has no major structural defects, but is obviously used. Excellent means the toy is in prime condition. It may have been played with, but if it was, it was kept in great shape and is bright, shiny, and colorful.

Buyer Beware! Recently, a catalogue offered a copy of the "Little Red Riding Hood" tea set. It is tin-litho and is a duplicate of the oldest "Little Red Riding Hood" set produced by Ohio Art in the 1920s. The duplicate is made in China and the advertiser proudly proclaims that it "still bears some little scratches like the old pieces." The inside of the teacups on the copies are bright shiny tin-colored, while the originals are likely to be quite worn and discolored inside. The teapot is much taller as well. No large round tray was produced by the imitators. Although the duplicates do not say Ohio Art, an inexperienced collector could be very easily fooled. No other duplicates of other Ohio Art toys are known of at this time, but it is important to beware of copies. Look for logos.

Photograph of Made in China "Little Red Riding Hood" set. *Courtesy of Sharon Lazane.*

DATING

A major advantage of collecting Ohio Art toys is the relative ease of dating pieces. Unlike other companies that produced the same lithography on their toys for many years, and sometimes even decades, Ohio Art changed its lithography frequently which makes it easier to date pieces. Moreover, Ohio Art changed its logo relatively frequently. The following chart depicts the sequential logos with corresponding dates:

1959-62 logo

1963-71 logo

Early Ohio Art logos. Used throughout the teens, 1920s, and 1930s.

1972-78 logo

1945-58 logo

1979-83 logo

1987 logo

1984-86 logo

At some point, Ohio Art began date coding their products. The simple code works as follows:

1. First numbers signify product identification.
2. The letter is for the month that production of the toy began: A is for January, B is for February, C is for March, etc.
3. The next number is for the day of the month.
4. The last two numbers stand for the year.

For example: 23-C179 stands for a toy iron, numbered 230 in the catalogue, that was first produced on March 1, 1979. Sometimes the code does not match up with the logo as it took several months to effect the changeover.

CHAPTER II: THE EARLY YEARS

CUPID AWAKE, CUPID ASLEEP

Ohio Art produced a variety of frames containing photos and prints. Although the majority of the frames were oval, the company produced straight-sided frames as well. The most popular, and surely the most successful were the framed Cupid Awake, Cupid Asleep prints which represent the first major success of the Ohio Art Company. The photographs were the work of M. B. Parkinson and were the idea of publisher and entrepreneur, Taber-Prang. The model was a young girl by the name of Josephine Anderson. In 1908, to advance the popularity of his oval frames, Mr. Winzeler began including prints within the frames. By 1910, he was producing 20,000 framed prints per day with the most popular being the Cupid prints. Mr. Winzeler offered Taber-Prang $100,000 for the exclusive rights to the Cupids. Although this was a very considerable sum of money in the early 1900s, Mr. Prang refused. Ohio Art sold 50 million of the Cupids in 1910. Considering the national population at the time was 92 million, this is an impressive figure indeed! Ohio Art continued to produce the framed Cupids through 1934. An ironic note: In 1938, after Ohio Art ceased production of the Cupid Asleep and Cupid Awake photos, Taber-Prang went into bankruptcy and Mr. Winzeler's son, Howie, purchased the rights to the photos for ten dollars!

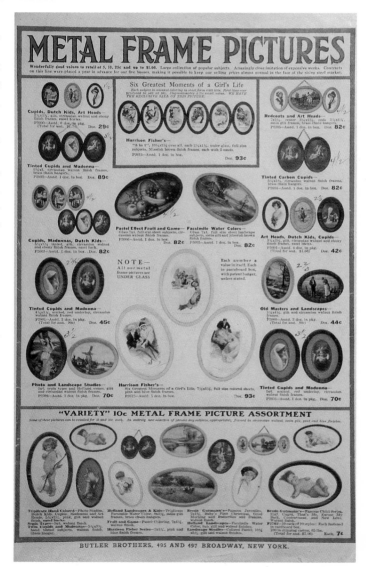

Early 1900s catalogue page showing early pictures sold in oval metal frames.

Early 1900s catalogue page displaying the oval framed "Cupid Awake" and "Cupid Asleep" mounted on a free-standing base.

No. 33—Easel Cupids and Madonnas

Frame stands 5 inches high, finished in Moorish brown. Three most popular selling subjects, Cupid Awake, Cupid Asleep and Madonna. Frames are equally well suited for photographs or post card size pictures which adds to their popularity.

Packed 2 in carton, ½ gross to the case. Weight per gross 38 lbs.

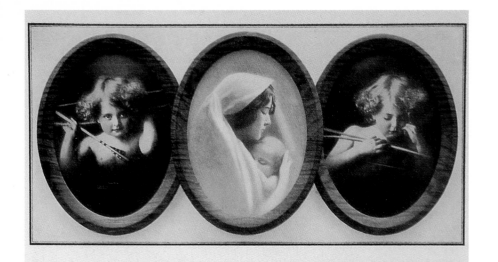

"Cupid Awake" and "Cupid Asleep." Produced between 1908 and 1935. $50-$65 for the set.

One of the early pictures mounted in an oval metal frame. Early 1900s. 5" high. *Courtesy of John Walters.* $30-$50.

Early 1900s catalogue page showing the famous "Cupid Awake" and "Cupid Asleep" as well as other popular works of art including "The Reapers." All of these reproductions were marketed in Ohio Art frames.

Some wood-grained metal oval frames. Early 1900s. 4-5". $10-$20.

TEA FOR TWO

Childrens' tea sets are adorable. They are such a symbol of old-fashioned girlhood and so difficult to find in their entirety.

Tin lithographed tea sets were one of the Ohio Art specialties. From the early 1920s until production ceased in 1983, Ohio Art manufactured hundreds of tea sets with a huge number of themes and patterns. Ohio Art was the most prolific producer of tea sets in America. The company had a separate tea set division turning out an incredible variety of designs, some of which were repeated for several years. It is interesting to note how a theme such as "Little Red Riding Hood" transformed over the years reflecting sociological changes over the decades from the subtle, intricate, idealized realism of the early 1900s to the cartoon-like over-brightness of the 1960s.

Tea sets were produced in a range of sizes. The older sets generally had straight-sided cups and a straight-sided coffee pot with a lid, bringing to mind Victorian Chocolate sets. Often the trays were round. In the late 1930s, cups and pots became rounded to a shape generally associated with tea service. For the purposes of this book, we will refer to the following most popular sizes:

Hummer: Five-piece set including one cup, one saucer, a tea-pot with lid, and a round plate used as a tray.
Seven-piece set: Two cups, two saucers, round plate tray, and straight-sided coffee pot with lid.
Nine-piece set: Two cups, two saucers, two dinner plates, coffee pot with lid, and tray.
Eleven-piece set: Two cups, two saucers, two dinner plates, teapot with lid, creamer, sugar, and tray.
Seventeen-piece set: Four cups, four saucers, four plates, teapot with lid, creamer, sugar, and tray.
Thirty-one-piece set: Six cups, six saucers, six plates, six butter plates, tea-pot with lid, creamer, sugar, cake plate with lid, and rectangular tray.

To acquire a set, mint-in-the box is a real find. It is also very expensive. More often than not, individual pieces are accumulated until an entire set is put together. Because water was the companion play item for tea sets, rust is very common. Look for pieces without rust. If there is some, it is better if it is on the inside rather than disturbing the exterior lithography. Mint-in-the-box is particularly nice in the case of tea sets. Mint-in-the-box insures that all of the pieces are from the same dye lot and represents a complete toy. With the lid removed, the box enhances display.

Although we do not have proof, after careful consideration and comparing notes, our best guess is that the first tea set produced by Ohio Art was the "Girl on the Swing" produced in 1918. It is probable that the company borrowed the dies for the set as the size of the cup and teapot is much larger than the sets produced in 1920.

The "Girl on the Swing" is utterly charming. It has the same soft, lyrical quality of many of the framed Ohio Art pictures. The litho is gorgeous from the shading of the bricks along the garden wall to the adorable smile of the little girl dressed in a soft rosy red that contrasts with the brick red of the swing and the rim circled with upper case ABCs in clear block letters.

"Girl on a Swing" tea set. Produced from 1918 through the early 1920s. Seven-piece set: $250-$450.

The "Kittens" is another early ABC set said to have been first displayed by H. S. Winzeler at the 1920 New York Toy Fair.

"ABC Kittens" tea set. Produced around 1920. Production continued through the late 1920s. Five-piece set: $200-$350.

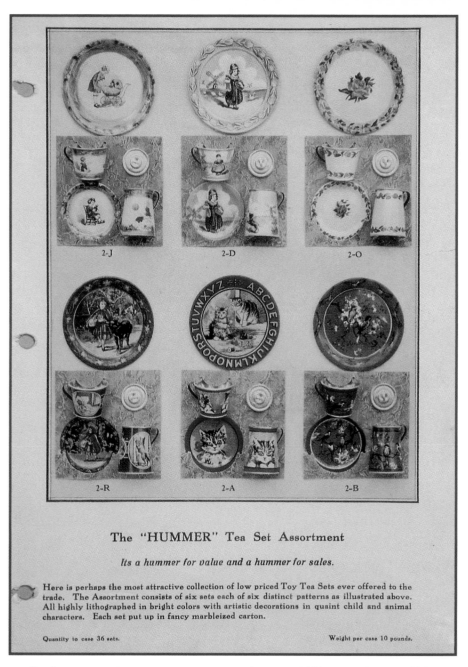

Catalogue page showing tea sets produced from late 1920s through early 1930s.

A collector recently sent us a photograph of a Peter Rabbit set we had never seen before. It is clearly an early set and we have not been able to locate this set in any catalogue. That is one of the enjoyable things about collecting. Who knows what other secrets are out there waiting to be discovered?

Adorable and rare early "Peter Rabbit" tea set. This set must have been produced in small quantities; even individual pieces are rarely found. *Courtesy of Susan L. Phillips.* Nine-piece set with tray: $300-$500.

The first tea sets produced after the ABC "Girl on a Swing" and "Kittens" sets were the "Little Red Riding Hood," "Little Girl at Play," "Delft Blue," "Butterfly," "Bluebirds," and "Roses" sets. These sets were advertised in late 1920s catalogues and continued to appear throughout the 1930s although there were some dramatic modifications to the "Little Red Riding Hood" set in the early 1930s. Some of these early sets were advertised and sold right through the end of the 1930s.

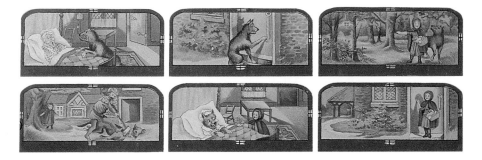

Prototype drawings for the first "Little Red Riding Hood" tea set produced by Ohio Art in the 1920s. *Courtesy of the Ohio Art Company.*

A close-up showing the fabulous lithography on the first "Little Red Riding Hood" tray. Note the realism of the wolf.

"Little Red Riding Hood" tea set. Produced from the late 1920s through approximately 1932. Nine-piece set: $225-$375.

A close-up of the teapot showing the wolf knocking at Grandma's door. The cup shows the wolf in bed wearing granny's cap and gown.

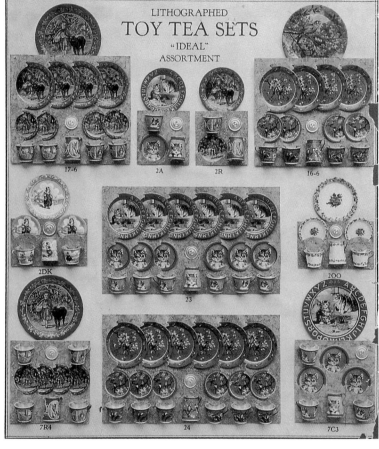

1928 catalogue page showing five of the earliest tea sets. It is likely that the "Girl on a Swing" set was no longer available by 1928.

"Delft Blue" tea set. 1929-1931. Eight-piece set: $125-$175.

"Early Rose" tea set. 1929-early 1930s. Eight-piece set: $135-$175.

"Little Girl at Play" tea set. 1929-early 1930s. Eight-piece set: $150-$250.

"Bluebirds and Blossoms" tea set. 1929-early 1930s. Eight-piece set: $135-$175.

OTHER TOYS

Other early toys include a train, sandtoys, an Easter cart, some little tin-litho wagons, possibly doll-size, and an alarm clock. The litho on these 1920s toys is especially charming and complex, incorporating the qualities found on late Victorian lithographed toys.

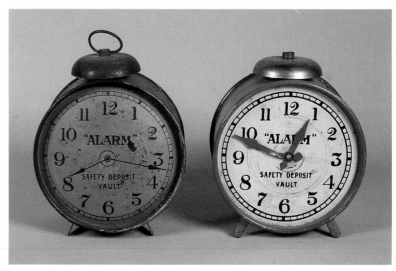

"Snowman and Little Girl" snow shovel. Late 1920s. The lithography has the same lyrical, soft quality found on the "Girl on a Swing" tea set. *Courtesy of Jim Gilcher*. $80-$150.

Early alarm clock. 1928. *Courtesy of Jim Gilcher*. $35-$65.

Left: Pretty enough to frame! First known catalogue sheet displaying toys produced in the late 1920s.

ABC Wagon. 1928. Courtesy of Jim Gilcher. $125-$250.

Two views of this charming Nursery Rhyme Wagon. 1928. $85-$135.

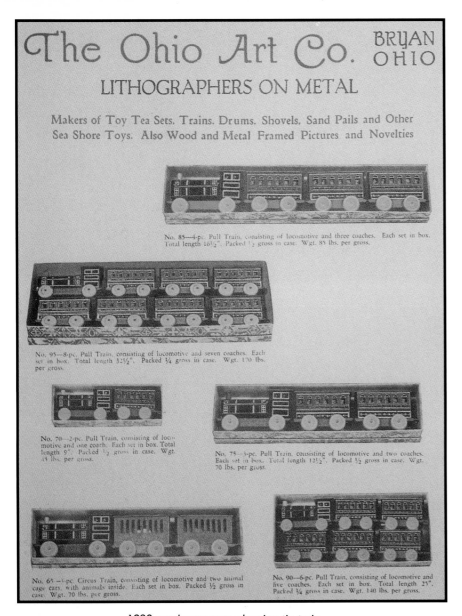

1928 catalogue page showing tin train.

Tin Train. 1928-1930s. *Courtesy of Jim Gilcher*. $350-$600.

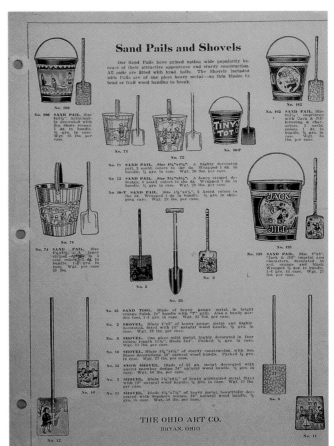

Early sand toy sheet.

Very rare early Easter toy with this charming little poem inscribed:

> Says Little Mister Rabbit
> While sitting on his tail
> I lay fresh eggs each morning
> And sell them by the pail.

This toy originally had two little buckets to hold goodies—one for each hand. *Courtesy of Jim Gilcher.* $200-$325.

Early "Tiny Tot" sand pail. *Courtesy of Jim Gilcher.* $120-$150.

Early sand pail. Late 1920s. *Courtesy of Jim Gilcher.* $100-$145.

Bunny Cart. Late 1920s. *Courtesy of Jim Gilcher*. $125-$250.

Close-up of back of Bunny Cart showing
colorful lithography.

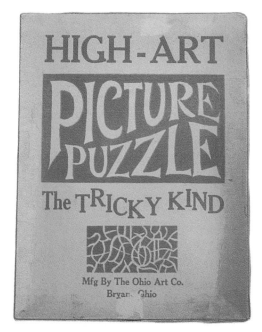

High Art Picture Puzzle, "The Cherry Orchard." 1920s. *Courtesy of Jim Gilcher.* $75-$200.

Sandlift. 1929-35. *Courtesy of Jim Gilcher.* $85-$145.

WALT DISNEY

The 1930s saw the flowering of Ohio Art Toys. The company tooled up for its large tea set production which enjoyed immense popularity. In addition to tea sets, Ohio Art also produced a huge variety of sandtoys including pails, sieves, and whole boxed "fun at the seashore" kits which included shovels and molds.

From the perspective of collectors, the single most important Ohio Art development of the 1930s was the licensing of the Walt Disney characters. By 1939, Kay Kamen Ltd., Walt Disney's licenser from 1933 on, had issued about 100 licenses to leading American manufacturers. However, no American company can rival Ohio Art for the gorgeous colors and lush lithography of its 1930s Disney toys. It's hard to beat the charms of the "Three Little Pigs" sand pail, "Mickey's Helpmate" tea set, and the "Mickey Plays the Sax" watering can.

"Mickey's Helpmate" tea set. This is the red-bordered set, mint-in-the box. The set came in a variety of sizes; the smallest one is pictured here. The larger sets included plates and sometimes a round or rectangular tray. *Courtesy of Bob Bernabe.* Smallest set MIB: $550. Larger set MIB: $750.

"Mickey's Helpmate" tea set. 1932-33. Copyright Walt Disney. This is the set with the orange border. *Courtesy of Doug and Pat Wengel.* Eight-piece set: $285-$425.

Close-up of wonderful lithography on "Mickey's Helpmate" plate.

Round and rectangular trays from "Mickey's Helpmate" set. *Courtesy of Bob Bernabe.*

Ohio Art first licensed with Walt Disney in the early 1930s. It is easy to date the 1930s Disney toys based on the changing licensing data that marked all Disney pieces. The following is a list of the dates and corresponding marks:

Between 1931 and 1933   "Walter E. Disney,"
                                    "Walt E. Disney,"
                                    "Copyright W.D.
1934-November 1939   "Walt Disney Enterprises"
December 1939-on   "Walt Disney Productions"

There are a group of toys that fall into a category called "changeover" toys that were produced in 1939-1940—right at the point that Disney changed from W.D.E. to Walt Disney Productions. For more exact dating, it should be noted that in the USA, Mickey Mouse's eyes went from wall-eyed to pie-eyed around 1933. When Donald Duck was first introduced in 1934, he had a long bill which appeared shorter in 1936.

OTHER ARTISTS

Walt Disney was not the only artist who produced designs for the Ohio Art Company. Artist Fern Bisel Peat, a well-known children's artist of the 1920s and 1930s, contributed many fabulous designs to the toys of Ohio Art. She illustrated more than 100 children's books, painted murals for public buildings and was art director for the children's magazine *Children's Play Mate*, from 1933 through 1953. As was true of many women illustrators of this period, her name remained in relative obscurity compared to famous male illustrators of the period. In addition to her books, puzzles, and paper dolls, she also decorated tea sets, sand buckets and tops for Ohio Art. These pieces are highly prized by today's collector.

Fern Bisel Peat illustration, "The Gingham Dog and the Calico Cat." 1930

"The Gingham Dog and Calico Cat" tea set. Fern Bisel Peat. 1930s. *Courtesy of Jim Gilcher*. Eight-piece set: $135-$225.

Fern Bisel Peat illustration, Girl with Fruit. 1930.

One of the artists who worked under the direction of Ms. Peat at *Children's Play Mate* magazine was Elaine Ends Hileman. Her work is also seen on several Ohio Art pieces. Other designers include Beatrice H.K. Benjamin, Ruth Sterling (RS), and Ruth Newton of the "Chubby Cubs" books. Jim Gilcher recently found another piece marked M.B.B.; the artist who belongs to these initials remains a mystery. Post World War II, the chief designer for Ohio Art has been Don Dean. He has signed at least four items.

TEA SETS

The tea sets of the 1930s were fabulous. Disney, Fern Bisel Peat, Elaine Ends Hileman, and many other designers contributed to the beauty of these sets. Some designs were continued from the late 1920s into the early 1930s.

Fern Bisel Peat illustration, "Peter Rabbit." 1943.

"Bunny Birthday" tea set. Late 1930s. Fern Bisel Peat. *Courtesy of Jim Gilcher.* Seven-piece set: $90-$150.

"Cinderella" tea set. Late 1930s. Fern Bisel Peat. Eleven-piece set: $135-$175.

"Humpty-Dumpty" tea set. Late 1930s. Fern Bisel Peat. This set came in several combinations. The tray of one set had Humpty-Dumpty falling off the wall, and one had him perched on the wall. All the other pieces have cute little birdies on a red gingham background. Some sets have the older straight-sided cups while others have rounded cups. *Courtesy of Jim Gilcher.* Eight-piece set: $75-$135.

"Little Maid with Flowers" tea set. Late 1930s through mid-1940s. Fern Bisel Peat.

Left: "Little Bo Peep" tea set. Late 1930s through mid-1940s. Fern Bisel Peat. Seven-piece set: $125-$175.

Left: "Little Red Riding Hood," tea set that first appeared in 1931 catalogue. Note the change on the tray, plate, and saucer. Note also that the flower pattern is broken into four segments instead of the consecutive ring. The cup and teapot remain the same as the earlier set. *Courtesy of Jim Gilcher.* Nine-piece set: $150-$225.

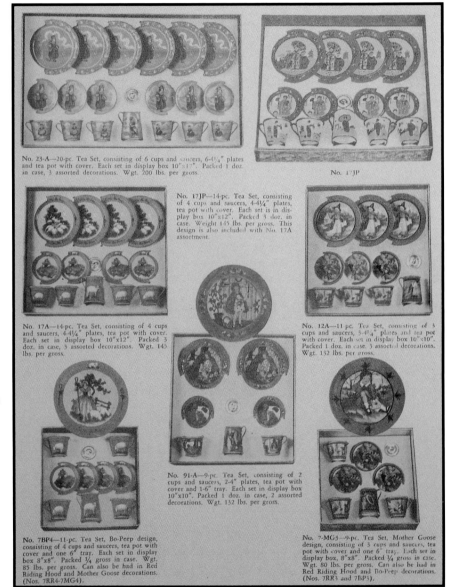

No. 23-A—20-pc. Tea Set, consisting of 6 cups and saucers, 6-4¼" plates and tea pot with cover. Each set in display box 10"x12". Packed 1 doz. in case, 3 assorted decorations. Wgt. 200 lbs. per gross.

No. 17JP

No. 17JP—14-pc. Tea Set, consisting of 4 cups and saucers, 4-4¼" plates, tea pot with cover. Each set is in display box 10"x12". Packed 3 doz. in case. Weight 145 lbs. per gross. This design is also included with No. 17A assortment.

No. 17A—14-pc. Tea Set, consisting of 4 cups and saucers, 4-4¼" plates, tea pot with cover. Each set in display box 10"x12". Packed 3 doz. in case, 3 assorted decorations. Wgt. 145 lbs. per gross.

No. 12A—11-pc. Tea Set, consisting of 3 cups and saucers, 3-4¼" plates and tea pot with cover. Each set in display box 10"x10". Packed 1 doz. in case, 3 assorted decorations. Wgt. 132 lbs. per gross.

No. 91-A—9-pc. Tea Set, consisting of 2 cups and saucers, 2-4" plates, tea pot with cover and 1-6" tray. Each set in display box 10"x10". Packed 1 doz. in case, 2 assorted decorations. Wgt. 132 lbs. per gross.

No. 7BP4—11-pc. Tea Set, Bo-Peep design, consisting of 4 cups and saucers, tea pot with cover and one 6" tray. Each set in display box 8"x8". Packed ¼ gross in case. Wgt. 85 lbs. per gross. Can also be had in Red Riding Hood and Mother Goose decorations. (Nos. 7RR4-7MG4).

No. 7-MG3—9-pc. Tea Set, Mother Goose design, consisting of 3 cups and saucers, tea pot with cover and one 6" tray. Each set in display box 8"x8". Packed ¼ gross in case. Wgt. 80 lbs. per gross. Can also be had in Red Riding Hood and Bo-Peep decorations. (Nos. 7RR3 and 7BP3).

Early 1930s catalogue page showing newly introduced dish sets. Note that some remains of 1920s inventory are included.

"Japanese Beauties" tea set. Early 1930s. Eight-piece set: $135-$185.

"Butterflies" tea set. This exquisite and hard-to-find tea set was produced in the early 1930s. Nine-piece set: $150-$275.

"Mother Goose" tea set. Early 1930s. "Mother Goose" is on the plates; "Jack and Jill" appear on the saucers; "Hey Diddle-Diddle" is on the teapot; and various nursery rhymes are depicted on the cups. The large charming circular tray is rimmed with stars depicting Jack after his fall. It appears very different than the rest of the set. Nine-piece set: $150-$275.

"Little Bo Peep" tea set. Early 1930s. Note the rhyme is written on the various pieces. The teapot is especially charming. Nine-piece: $150-$275.

1933-34 catalogue page. Note the introduction of two very important teasets: "Mickey's Helpmates" and "Fairies."

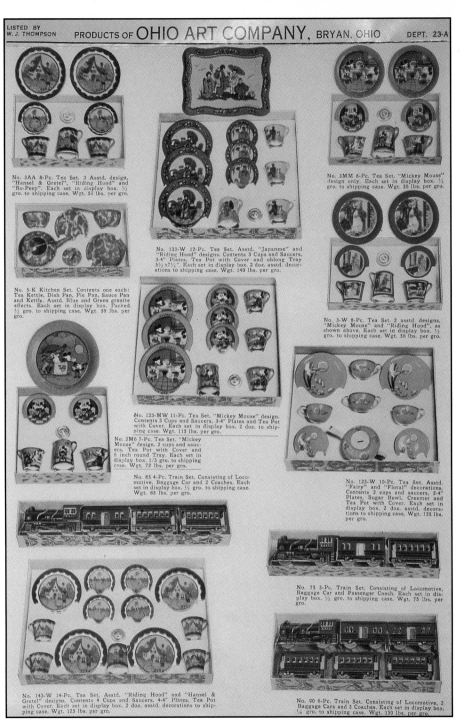

"Hummer"—5-pc. Tea Set consisting of cup and saucer, 4¼" plate and tea pot with cover. Each set in 5"x5" display box. Packed 3 doz. sets in case, 6 assorted decorations, weight 40 lbs. per gross.

No. 2-AA—7-pc. Tea Set consisting of 2 cups and saucers, tea pot with cover and 4¼" tray. Each set in display box 4¼"x 7½". Packed ½ gross in case, 3 assorted decorations. Wgt. 46 lbs. per gross.

No. 3-BP—8-pc. Tea Set, Bo-Peep design, consisting of 2 cups and saucers, 2 4¼" plates and tea pot with cover. Each set in display box 5"x8½". Packed ½ gross in case. Wgt. 56 lbs. per gross.

No. 3-MG—8-pc. Tea Set, Mother Goose design, consisting of 2 cups and saucers, 2-4¼" plates, tea pot with cover. Each set in display box 5"x8½". Packed ½ gross in case. Wgt. 56 lbs. per gross.

No. 3R6—7-pc. Tea Set, Red Riding Hood design, consisting of 2 cups and saucers. Tea Pot with cover, and one 6" tray. Each set in display box 7"x 7". Packed ⅓ gross in case. Wgt. 72 lbs. per gross. Can also be had in Bo-Peep and Mother Goose decorations (Nos. 3BP6 and 3MG6).

No. 3-RR—8-pc. Tea Set, Red Riding Hood design, consisting of 2 cups and saucers, 2 4¼" plates and tea pot with cover. Each set in display box 5"x8½". Packed ½ gross in case. Wgt. 56 lbs. per gross.

No. 3-JP—8-pc. Tea Set, Japanese design consisting of 2 cups and saucers, 2-4¼" plates, tea pot with cover. Each set in display box 5"x8½". Packed ½ gross in case. Wgt. 56 lbs. per gross.

Early 1930s catalogue page.

"Three Little Pigs" tea set. 1934. Walt Disney Enterprises. Nine-piece set: $225-$325.

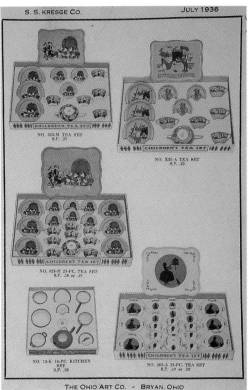

1936-37 catalogue showing "Mickey Presenting Minnie" teaset.

"Mickey Presenting Minnie" tea set. 1936-37. Walt Disney Enterprises. This set shows many characters including the long-billed Donald. *Courtesy of Doug and Pat Wengel.* Nine-piece set: $300-$425.

"Clara Clack" tea set. 1939. Walt Disney Enterprises. *Courtesy of Doug and Pat Wengel.* Nine-piece set: $225-$375.

Tray from
the "Clara Clack" tea set.

"Fairies" tea set. 1933-35. This utterly charming set has a soft, beautiful blue background which contributes to its desirability. Nine-piece set: $150-$300.

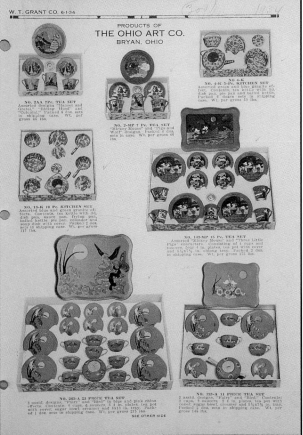

"Snow White and the Seven Dwarfs" tea set. 1938. Walt Disney Enterprises. This was the first Ohio Art teaset produced connected to a full-length animated Disney feature. Nine-piece set: $260-$400.

1934 catalogue page showing a variety of tea sets.

"Pink Bird" teaset. 1937. *Courtesy of Jim Gilcher.* Nine-piece set: $135-$235.

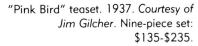

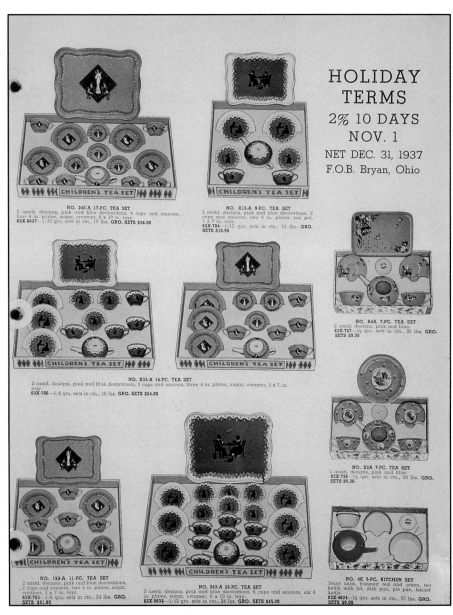

NO. 243-A 17-PC. TEA SET
2 asstd. designs, pink and blue decorations, 4 cups and saucers, four 4 in. plates, sugar, creamer, 8 x 10 in. tray.
63X-8637—1/12 gro. sets in ctn., 19 lbs. GRO. SETS $36.00

NO. X12-A 9-PC. TEA SET
2 asstd. designs, pink and blue decorations, 2 cups and saucers, two 4 in. plates, tea pot, 5 x 7 in. tray.
63X-704—1/12 gro. sets in ctn., 11 lbs. GRO. SETS $19.80

NO. B4A 7-PC. TEA SET
2 asstd. designs, pink and blue.
63X-727—⅓ gro. sets in ctn., 32 lbs. GRO. SETS $9.36

NO. X35-A 14-PC. TEA SET
2 asstd. designs, pink and blue. decorations, 3 cups and saucers, three 4 in. plates, sugar, creamer, 5 x 7 in. tray.
63X-706—1/6 gro. sets in ctn., 26 lbs. GRO. SETS $24.00

NO. B3A 7-PC. TEA SET
2 asstd. designs, pink and blue.
63X-726—⅓ gro. sets in ctn., 30 lbs. GRO. SETS $9.36

NO. 123-A 11-PC. TEA SET
2 asstd. designs, pink and blue decorations, 2 cups and saucers, two 4 in. plates, sugar, creamer, 5 x 7 in. tray.
63X-705—1/6 gro. sets in ctn., 24 lbs. GRO. SETS $21.60

NO. 263-A 23-PC. TEA SET
2 asstd. designs, pink and blue decorations, 6 cups and saucers, six 4 in. plates, sugar, creamer, 8 x 10 in. tray.
63X-8632—1/12 gro. sets in ctn., 24 lbs. GRO. SETS $45.00

NO. 4K 5-PC. KITCHEN SET
Ivory units, trimmed red and green, tea kettle with lid, dish pan, pie plan, boiled kettle.
62X-4624—⅓ gro. sets in ctn., 25 lbs. GRO. SETS $9.00

1937 catalogue page showing silhouette teaset.

"Silhouette" tray.

BY THE SEA

Sand pails are a perfect medium for lithography. The curved shape is amenable to a sequential series of events. The shape is simple and even, making it an ideal canvas for design. The association with childhood days at the seashore gives sand pails a lovely emotional appeal.

The Disney sand pails are nothing short of exquisite. There is simply no Disney litho produced by any company that compares with that produced by Ohio Art during the 1930s. Some pails depicted an entire plot with various characters acting out a scene. The Ohio Art Walt Disney lithography was not confined to Mickey and Minnie Mouse. Several scenes from

Disney features such as Snow White and the Three Little Pigs were the subjects of Ohio Art lithography. More rarely visited characters such as Clara Clack were also subject material.

Fern Bisel Peat designed several wonderful sand pails during the 1930s. Other designs from the 1930s are also sought after by collectors.

"Mickey and Minnie as Pirates" sand pail. 1931-33. Copyright Walt Disney. 6" tall. *Courtesy of Bob Bernabe.* $300-$500.

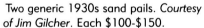

Two generic 1930s sand pails. *Courtesy of Jim Gilcher.* Each $100-$150.

Two views of "Mickey Magician" sand pail. Copyright Walt Disney. Circa 1933. 3 1/4" tall. *Courtesy of Doug and Pat Wengel.* $200-$400.

Two views of "Mickey and Minnie at the Beach" sand pail. Walt Disney Enterprises. Mid-1930s. 4 1/4" tall. *Courtesy of Doug and Pat Wengel.* $200-$400.

Two views of "Farmcrest" sand pail. Walt Disney Enterprises. Mid-1930s. 4 1/2" tall. Sometimes a business would contract with Ohio Art to have their name printed on a pail as an endorsement for their company or product. "Farmcrest" was most likely a dairy. The pail was also available without the endorsement. *Courtesy of Doug and Pat Wengel.* $225-$350.

"Mickey's Garden" sand pail. Mid-1930s. Walt Disney Enterprises. This pail has a lip base which was rare for the Ohio Art Company. 5 1/2" tall. *Courtesy Bob Bernabe.* $350-$550.

Two views of "Atlantic City Automotive Mickey" sand pail. Walt Disney Enterprises. Mid-1930s. This pail came with and without "Atlantic City" written on it. 8" tall. *Courtesy of Doug and Pat Wengel.* Excellent condition, with "Atlantic City" $450-$800. Without "Atlantic City" $350-$700.

"Mickey Catching Fish" sand pail. Mid-1930s. Walt Disney Enterprises. 8" tall. *Courtesy Bob Bernabe.* $400-$800.

Two views of "Mickey and Minnie at Home" sand pail. Walt Disney Enterprises. Mid-1930s. This pail also came with and without "Atlantic City" written on the pail. Note Donald and Mickey on the back. 8" tall. *Courtesy of Doug and Pat Wengel*. Excellent, with "Atlantic City" $450-$800. Without "Atlantic City" $350-$700.

Two views of "Mickey Fishing" sand pail. Walt Disney Enterprise. Circa 1934. This is another of the rare platform pails produced by Ohio Art Company. 7" tall. *Courtesy of Doug and Pat Wengel.* $450-$850.

Two views of "Mickey's Band" sand pail. Walt Disney Enterprises. Mid-1930s. This wonderful pail features a large variety of characters and is difficult to find in good shape. 5 3/4" tall. *Courtesy of Doug and Pat Wengel.* $450-$750.

Two views embossed Disney sand pail. Walt Disney Enterprises. Mid-1930s. The characters and beach umbrella are embossed. 5 3/4" tall. *Courtesy of Doug and Pat Wengel.* $325-$650.

Two views
"Small Atlantic City" sand pail. Walt Disney Enterprises. Mid-1930s. This rarer pail was only made with the "Atlantic City" printing. 4 1/2" tall. *Courtesy of Doug and Pat Wengel.* $500-$750.

"Green Boat" sand pail. Walt Disney Enterprises. Mid-1930s. This pail features Mickey, Minnie, Donald, and Pluto. 5" tall. *Courtesy of Doug and Pat Wengel.* $300-$400.

Two views of "Three Little Pigs" sand pail. Walt Disney Enterprises. 1934. 6" tall. *Courtesy of Doug and Pat Wengel*. $275-$600.

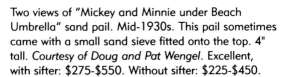

Two views of "Mickey and Minnie under Beach Umbrella" sand pail. Mid-1930s. This pail sometimes came with a small sand sieve fitted onto the top. 4" tall. *Courtesy of Doug and Pat Wengel*. Excellent, with sifter: $275-$550. Without sifter: $225-$450.

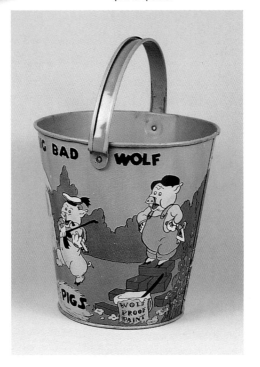

1937 catalogue page showing various sand toys.

1939 catalogue page showing various sand toys.

1938 catalogue page showing various sand toys.

"Snow White" medium sand pail. Walt Disney Enterprises. 1938. 5 1/2" tall. *Courtesy of Doug and Pat Wengel.* $225-$425.

"Snow White" small sand pail. Walt Disney Enterprises. 1938. 4" tall. *Courtesy of Doug and Pat Wengel.* $225-$325.

"Snow White" large sand pail. Walt Disney Enterprises. 1938. 8" tall. *Courtesy of Doug and Pat Wengel.* $300-$500.

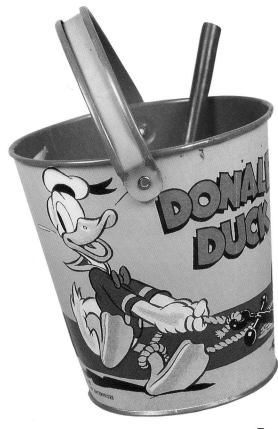

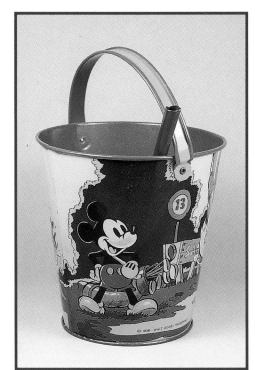

Two views "Golf" sand pail. Walt Disney Enterprises. 1938. 5 3/4" tall. *Courtesy of Doug and Pat Wengel.* $225-$550.

Two views "Donald, Mickey, and Minnie at Sea" sand pail. Walt Disney Enterprises. 1938. 4" tall. *Courtesy of Doug and Pat Wengel.* $175-$350.

Two views "Mickey's Lemonade Stand" sand pail. Walt Disney Enterprises. Mid-1930s. 2 3/4" tall. *Courtesy of Doug and Pat Wengel.* $175-$325.

"Treasure Island" sand pail. Walt Disney Enterprises. 1935. 5 1/2" tall. *Courtesy of Doug and Pat Wengel.* $275-$425.

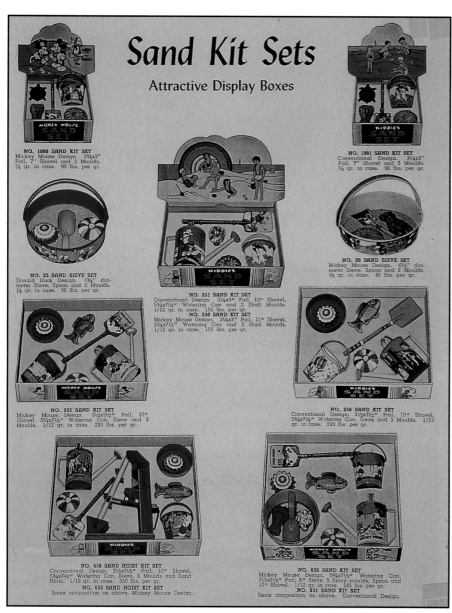

Sand Kit Sets

Attractive Display Boxes

NO. 1000 SAND KIT SET
Mickey Mouse Design. 3¾x3"
Pail, 7" Shovel and 3 Moulds.
¼ gr. in case. 90 lbs. per gr.

NO. 25 SAND SIEVE SET
Donald Duck Design. 7¾" diameter Sieve, Spoon and 2 Moulds.
½ gr. in case. 56 lbs. per gr.

NO. 1001 SAND KIT SET
Conventional Design. 3¾x3"
Pail, 7" Shovel and 3 Moulds.
¼ gr. in case. 90 lbs. per gr.

NO. 20 SAND SIEVE SET
Mickey Mouse Design. 6½" diameter Sieve, Spoon and 2 Moulds.
½ gr. in case. 45 lbs. per gr.

NO. 251 SAND KIT SET
Conventional Design. 3¾x3" Pail, 10" Shovel,
5¾x7½" Watering Can and 2 Shell Moulds.
1/12 gr. in case. 155 lbs. per gr.

NO. 250 SAND KIT SET
Mickey Mouse Design. 3¾x3" Pail, 10" Shovel,
5¾x7½" Watering Can and 2 Shell Moulds.
1/12 gr. in case. 155 lbs. per gr.

NO. 255 SAND KIT SET
Mickey Mouse Design. 3½x3½" Pail, 10"
Shovel, 5¾x7½" Watering Can, Sieve and 3
Moulds. 1/12 gr. in case. 230 lbs. per gr.

NO. 256 SAND KIT SET
Conventional Design. 3½x3½" Pail, 10" Shovel,
5¾x7½" Watering Can, Sieve and 3 Moulds. 1/12
gr. in case. 230 lbs. per gr.

NO. 656 SAND HOIST KIT SET
Conventional Design. 3½x3½" Pail, 10" Shovel,
5¾x7½" Watering Can, Sieve, 3 Moulds and Sand
Hoist. 1/12 gr. in case. 350 lbs. per gr.
NO. 655 SAND HOIST KIT SET
Same composition as above, Mickey Mouse Design.

NO. 650 SAND KIT SET
Mickey Mouse Design. 5¾x7½" Watering Can,
3½x3½" Pail, 6" Sieve, 5 fancy moulds, Spoon and
10" Shovel. 1/12 gr. in case. 245 lbs. per gr.
NO. 651 SAND KIT SET
Same composition as above. Conventional Design.

1939 catalogue page.

Five Walt Disney Enterprises sand shovels in progressive size order. Smallest shovel 2 1/2" tall by 1 1/2" wide. *Courtesy of Doug and Pat Wengel.* $90-$200. "Mickey and Minnie on the Beach" shovel. 3 1/4" tall by 2 3/8" wide. *Courtesy of Doug and Pat Wengel.* $135-$250. "Mickey and the Curling Wave" shovel. 3 5/8" tall by 3" wide. *Courtesy of Doug and Pat Wengel.* $150-$325. "Mickey Digging Sand" shovel. 5" tall by 4" wide. *Courtesy of Bob Bernabe.* $225-$400. "Mickey Testing Water" shovel. 7 1/4" wide by 6 1/4" tall. $275-$500.

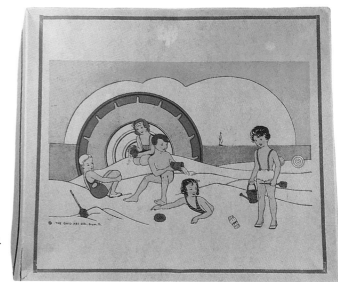

Box for sand set.

"Mickey Mouse Sand Set" in original generic box, No. 650. Walt Disney Enterprises. 1937-38. The set contains second-smallest shovel, "Donkey" watering can, "Mickey Playing Banjo" sifter, Donald pail, and several generic sand molds. *Courtesy of Doug and Pat Wengel.* Mint-in-Mint Box $1200. Mint-in-Excellent Box $1000.

Two views of Mickey Mouse sand set in original box. No. 1000. Walt Disney Enterprises. 1934 and 1937. Box is 7" by 7". This sand set came in three sizes. The smallest, pictured here, contains a "Mickey's Lemonade Stand" sand pail, a shovel, and three generic sand molds. *Courtesy of Doug and Pat Wengel.* Mint-in-mint box $1500. Mint-in-Excellent box $1350.

"Banjo Mickey" sand sifter. Walt Disney Enterprises. 1938. 6 1/4" in diameter. *Courtesy of Doug and Pat Wengel.* $125-$250.

"Mickey and Friends" sand sifter. Walt Disney Enterprises. Mid-1930s. Horace and Clarabell are not pictured on toys as frequently after 1938. 7 1/2" in diameter. *Courtesy of Doug and Pat Wengel.* $150-$300.

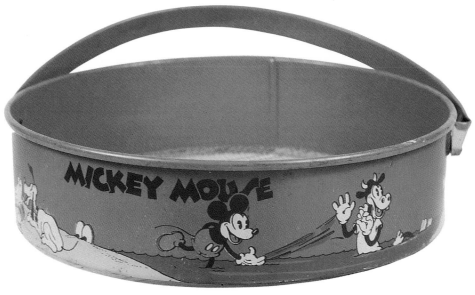

OTHER TOYS

Tops and drums were also popular in the 1920s. The artists who designed tea sets and sand toys lent their skills to the design of these toys as well.

"Noah's Ark" drum. Early 1930s. *Courtesy of Jim Gilcher.* $90-$135.

1936 catalogue page showing tops, drums, and a snow shovel.

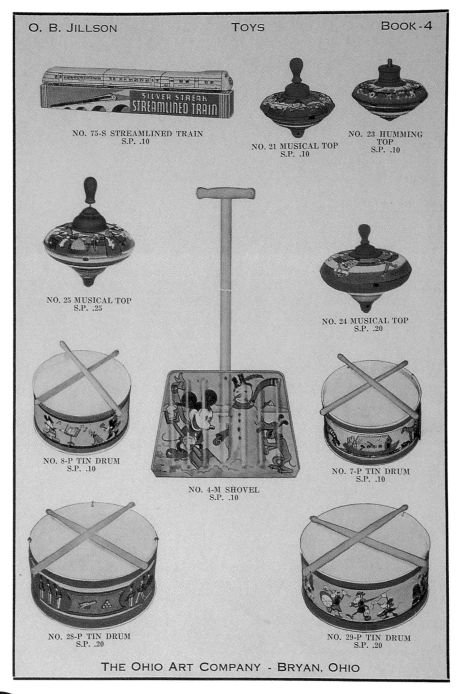

"Soldier" drum. 1930s. *Courtesy of Jim Gilcher.* $100-$140.

"Mickey Mouse Musician" drum. Copyright Walt Disney. 1933. This is one of the earliest Disney drums. It has a metal drum head. 6 1/2" diameter. *Courtesy of Doug and Pat Wengel.* $225-$450.

Two views "Mickey Mouse Parade" drum. Copyright Walt Disney. 1933. This is another early Disney drum with a metal head. 9" diameter. *Courtesy of Doug and Pat Wengel.* $275-$500.

Top right and bottom: Mickey Mouse Drum. Walt Disney Enterprises. Mid-1930s. This charming drum with metal heads includes a large assortment of Disney characters including Clara Cluck. The background bears a resemblance to Central Park in New York City. *Courtesy of Doug and Pat Wengel.* $300-$425.

Two views of a different and slightly later "Mickey Mouse Parade" drum. Walt Disney Enterprises. Mid-1930s. Metal head. 6 1/4" in diameter. *Courtesy Doug and Pat Wengel.* $250-$375.

Two views of Mickey Mouse Band Drum. Circa 1935. This cloth head drum is not marked, but it attributed to the Ohio Art Company by some collectors. 12 3/4" in diameter. *Courtesy of Doug and Pat Wengel*. $225-$325.

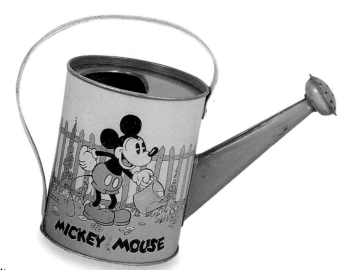

"Mickey in Garden" watering can. Walt Disney Enterprises. 4 1/2" tall. Courtesy of Doug and Pat Wengel. $225-$400.

"Mickey with Chickens" watering can. Walt Disney Enterprises. Approximately 4 1/2" tall. Courtesy of Doug and Pat Wengel. $225-$400.

1935 catalogue page showing watering cans and shovel.

NO. 33—WATERING CAN
Mickey Mouse design. Overall size 6½x8½". 1-3 gro. in case. Wt. per gro. 60 lbs.

NO. 32—WATERING CAN
Conventional. Overall size 6½x8½". 1-3 gro. in case. Wt. per gro. 60 lbs.

NO. 34—WATERING CAN
Three Pigs design. Overall 6½x8½". 1-3 gro. in case. Wt. per gro. 60 lbs.

NO. 36—WATERING CAN
English Garden. Overall size 9x11". 1 doz. in case. Wt. per gro 150 lbs.

NO. 38—WATERING CAN
Mickey Mouse. Overall size 9x11". 1 doz. in case. Wt. per gro. 150 lbs.

NO. 11-M—SHOVEL
6½x7½" blade, 24" red handle, Mickey Mouse design. ½ gro. in case. Wt. per gro. 90 lbs.

NO. 2-M—SHOVEL
4¾x5" blade 15" red handle, Mickey Mouse design. ½ gro. in case. Wt. per gro. 30 lbs.

NO. 25—SAND SIEVE SET
7¾" Mickey Mouse Sieve, 2 moulds and shovel. ½ gro. in case. Wt. per gro. 55 lbs.

NO. 100S—SAND KIT SET
Three Pigs design, 3¼x3" pail, 7" shovel, 3 moulds, ¼ gro. in case. Wt. per gro. 90 lbs.

NO. 250—SAND KIT SET
Mickey Mouse design, 3¼x3" pail, 10" shovel, 5¾x7½" watering can, 2 shell moulds, 1 doz. in case. Wt. per gro. 155 lbs.

NO. 258
Three Little Pigs design. Otherwise same as above.

NO. 1000—SAND KIT SET
Mickey Mouse design, 3¼x3" pail, 7" shovel and 3 moulds, ¼ gro. in case. Wt. per gro. 90 lbs.

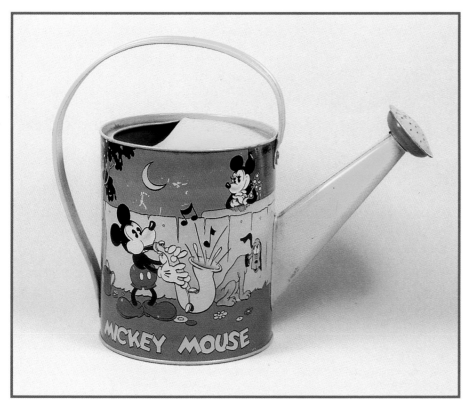

"Mickey Mouse Blowing the Sax" watering can. Walt Disney Enterprises. Mid-1930s. This is one of the best scenes on any of the Disney toys. It has lots of soul! 6" tall. *Courtesy of Doug and Pat Wengel.* $300-$450.

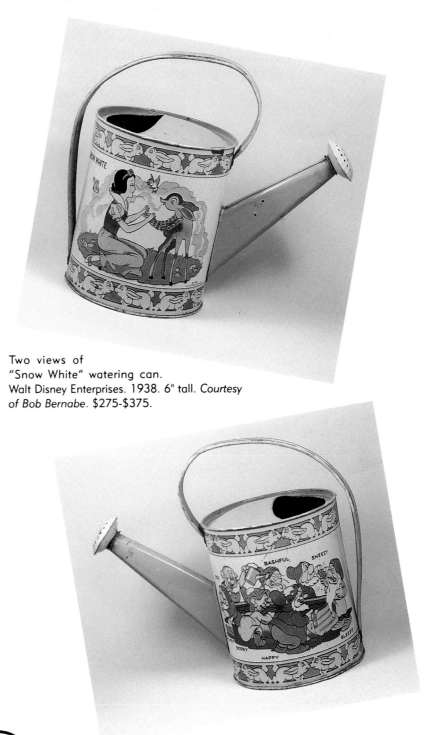

Two views of "Snow White" watering can. Walt Disney Enterprises. 1938. 6" tall. *Courtesy of Bob Bernabe.* $275-$375.

"Girl with Cap" sprinkling can. Late 1930s. Fern Bisel Peat. $75-$125.

"Girl in Garden" sprinkling can. Late 1930s through early 1940s. Fern Bisel Peat *Courtesy of Shirley Cox*. $75-$125.

Two views of "Donald and the Donkey" watering can. Walt Disney Enterprises. 1938. 4 1/2" tall. *Courtesy of Doug and Pat Wengel.* $200-$350.

"Mickey and Pluto with Snowman" snow shovel. Copyright Walt Disney. 1932-33. 9 1/2" wide at base. *Courtesy of Doug and Pat Wengel.* $450-$750.

"Donald Snowball Fight" snow shovel. Walt Disney Enterprises. 1935. 9 1/2" base. *Courtesy of Bob Bernabe.* $250-$500.

Mickey Mouse laundry set in the box and out of the box. Walt Disney Enterprises. 1935. This set also came with the "Three Little Pigs" design. *Courtesy of Doug and Pat Wengel.*

"Mickey Mouse" washer, mint-in-box. Walt Disney Enterprises. 1934. This washer also came without the wringer attached. *Courtesy Doug and Pat Wengel*. Mint-in-Box $1200. Without box $400-$600.

Box for washer.

"Mickey Mouse" washer without ringer. *Courtesy of Doug and Pat Wengel*.

"Seasaw" music box. 1939 to 1940s. 4 1/2" high. Another example of the lovely, soft lyrical lithography found on many of the early toys. *Courtesy of Michael Milson*. $60-$85.

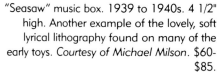

"Pirates" top. Late 1930s. Fern Bisel Peat. $70-$100.

"Cats" top. 1930s.
$55-$75.

Close-up of Fern Bisel Peat signature.

"Characters" top. 1930s.
Courtesy of Jim Gilcher.
$50-$70.

Sports lunch box.
1930s. *Courtesy of Jim Gilcher.* $45-$75.

Butterfly tray. 1930s. *Courtesy of the Ohio Art Company.* $40-$65.

Sandpump. 1937. *Courtesy of Jim Gilcher.* $75-$125.

1930s catalogue page showing "silhouettes" sold by the Ohio Art Company in the 1930s.

CHRISTMAS TREE DECORATIONS

Silver Ribbon Icicles, Colored Icicles, Alaska Snow, Artificial Snow, Ornament Hangers, Sparkling Floss, Etc.

Our especially designed machines enable us to produce Silver Ribbon Icicles with a glittering sheen not possible by any other method. New and improved display packages are added features which greatly stimulate sales. This is your most profitable line of Christmas decorations and should be prominently featured.

No. 43
Silver Ribbon Icicles

A large and attractive 4-color package containing 150—20-inch strands of lustrous icicles. Size of box 5"x8"x⅝". Packed 1 gross in case. Weight 22 lbs. per case.

No. 49
Silver Ribbon Icicles

An attractive new style carton, large and flashy. Contains 200—20-inch strands. Size of box 6"x9½"x⅝". Packed 1 gross in case. Weight 30 lbs. per case.

No. 44

Same style package as above, except smaller. Contains 100, 18-inch strands. Size of box 4½"x7½"x⅝" Weight 20 lbs. per case

No. 150
Colored Icicles

Mixed colors in an attractive carton size 4"x7"x⅝". Contains 65—20-inch strands. Packed 1 gross in case. Weight 15 lbs. per case
Can also be had in solid colors, Viz: No. 11, Red; No. 12 Green; No. 13, Gold; No. 16, Blue.
Each color packed solidly 1 gross in case

No. 42
Silver Ribbon Icicles

A beautiful 4-color package containing 150—18" strands. Size of box 4½"x-7½"x⅝". Packed 1 gross in case. Weight 24 lbs. per case

No. 42 1-2

Same package as above except that it contains 125—18-inch strands. Weight 22 lbs. per case.

No. 201
Rainbow Icicles

Due to the multi-color effect these rainbow icicles are exceptionally attractive. 4 doz. glassine containers in a specially designed display box. Size of display box 6¾"x10"x3½". Packed 1 gross in case. Weight 15 lbs. per case

NO. 48
Silver Ribbon Icicles

A big flashy package with wreath and candle design. Cut out effect shows contents thru openings. Contains 200—20-inch strands. Size of box 5½"x9"x⅝". Packed 1 gro. in case. Weight 30 lbs. per case

No. 10
Silver Ribbon Icicles

An attractive 4-color package to retail at 5c. Size of box 5"x5"x⅝". Contains 75—18-inch strands. Packed 1 gross in case. Weight 15 lbs. per case

No. 10 1-2

Same package as above except that it contains 60—18-inch strands.

Snap-On Ornament Hangings

A new and improved ornament hanger. Holds article securely to tree and is easily attached. See illustration at left. 40 hangers in box for 10c. Put up in especially designed counter display box contains 48 packages. Packed 1 gross to the case. Weight 10 lbs. per case.

No. 301
Crushed Stone

White crushed stone especially suited as ballast for miniature railroads, pathways for lawns, driveways, rockeries and other decorative purposes. A liberal sized package to retail at 10c. Packed ½ gross in a case. Weight 40 lbs. per case

No. 15
Artifical Snow

Very realistic and effective as floor covering for display windows, show cases, etc. Is fire-proof and has the glitter of genuine snow. Put up in securely sealed packages containing 4 oz. to retail at 10c.

No. 14

Same as above, 2 oz. package to retail at 5c.

No. 300
Artificial Grass

Extensively used for miniature lawns, gardens, golf courses, parks, etc. Also desirable for Easter decorations or anywhere the effect of real grass is desired. Put up in attractive packages 5½"x7"x2". Packed ½ gross in case. Weight 40 lbs. per case.

No. 101
Sure-On Ornament Hangers

Especially designed hanger with spring lock which securely holds ornament. A simple but effective hanger. Put up 50 in a box for 10c. 4 doz. boxes in a fancy counter display package. Packed 1 gross in a case. Weight 10 lbs. per case

No. 18
Alaska Snow

A combination of asbestos and mica, highly decorative and absolutely fire-proof. Alaska Snow is especially desirable for Christmas tree decorations. Size of box 4½"x6¾"x1". Packed 1 gross in case. Weight 25 lbs. per case

No. 20
Alaska Snow

For realistic effect, Alaska Snow is most desirable. It is light and fluffy and glistens like genuine snow, and being non-inflammable is free from fire hazard. A handsomely decorated carton 5½"x7"x1⅛". 1 gross in case. Weight 35 lbs. per case.

Two 1931 catalogue pages showing the line of Christmas decorations either produced or marketed by the Ohio Art Company.

Ohio Art continued to produce tea sets, sand toys, tops, and drums throughout the 1940s. Fern Bisel Peat and other artists continued to produce designs for Ohio Art toys. Some of the 1930s designs were continued into the 1940s as well.

Walt Disney Productions continued to design for Ohio Art in the 1940s. In the 1939 issue of *Playthings*, Ohio Art advertised, with great anticipation, the imminent availability of a Pinocchio catalogue. Due to licensing constraints, the Pinocchio toys were not actually available until 1940.

In 1942, Ohio Art stopped producing toys in order to turn out rocket parts and shell casings for World War II. The company continued to market existing inventory, and in 1945 resumed toy production.

"Pinocchio" tea set. Walt Disney Productions. 1940. Eleven-piece set: $225-$375.

1939 advertisement from *Playthings* magazine announcing Pinocchio products due to be released.

"Chef Donald" tea set. Walt Disney Productions. 1942. *Courtesy of Jim Gilcher.* Eleven-piece set: $175-$300.

"Puppies" tea set. Early 1940s. This set has the look of Fern Bisel Peat. *Courtesy of Shirley Cox.* Thirty-one piece set: $220-$325.

"Jack Be Nimble" plate from "Mother Goose" tea set. Early 1940s. Fern Bisel Peat. Seven-piece set: $90-$175.

"Ducky Bath Time" tea set. 1940s. *Courtesy of Jim Gilcher*. Thirty-one piece set: $220-$325.

"Girls as Kittens" tea set. Late 1930s-mid-1940s. Fern Bisel Peat. Thirty-one piece set: $250-$375.

"Mexican Boys" tea set. Beatrice Benjamin. Early 1940s. *Courtesy of Jim Gilcher*. Eleven-piece set: $135-$165.

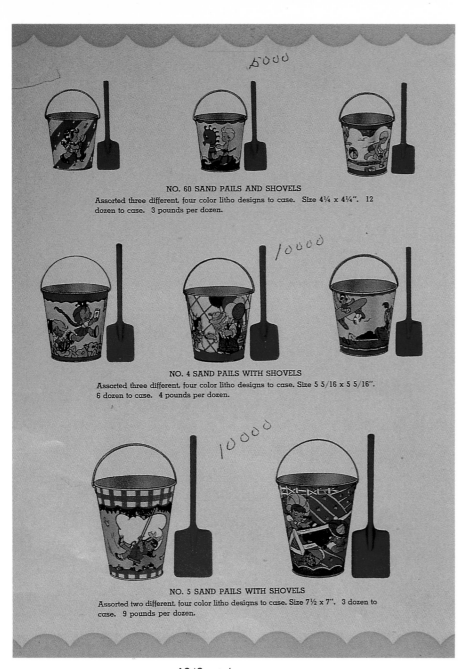

NO. 60 SAND PAILS AND SHOVELS
Assorted three different, four color litho designs to case. Size 4¼ x 4¼". 12 dozen to case. 3 pounds per dozen.

NO. 4 SAND PAILS WITH SHOVELS
Assorted three different, four color litho designs to case. Size 5 5/16 x 5 5/16". 6 dozen to case. 4 pounds per dozen.

NO. 5 SAND PAILS WITH SHOVELS
Assorted two different, four color litho designs to case. Size 7½ x 7". 3 dozen to case. 9 pounds per dozen.

1949 catalogue page.

Three 1940s sand pails.

"Polar Bears" sand pail. 1942. Fern Bisel Peat. Large pail. *Courtesy of Claudette Job.* $135-$175.

"Battleship Animals" sand pail. 1942. Small size. *Courtesy of Claudette Job*. $85-$135.

"Butterflies" sand pail. Late 1940s. 5" tall. *Courtesy of Doug and Pat Wengel*. $85-$135.

"Carousel" sand pail. No. 106. 1945-59. Fern Bisel Peat. 7 3/4" tall. *Courtesy of Doug and Pat Wengel*. $125-$160.

"Fisherman, Sailor Boy, and Pretty Girl" sand pail. 1940s. 8" tall. *Courtesy of Doug and Pat Wengel.* $120-$165.

"Ducky Bath" sand pail early 1940s. Elaine Ends Hileman. Small size. *Courtesy of Doug and Pat Wengel.* $95-$125.

"Off to the Beach" sand pail. 1940s-50s. Elaine Ends Hileman. 8" tall. *Courtesy of Doug and Pat Wengel.* $130-$160.

"Mexican Children" sand pail. Late 1940s-early 1950s. Small size. $75-$95.

"Donald Disneyland" Overland Candy Co., Chicago, Illinois, 1949, Walt Disney Productions. 2 2/3" tall. Not marked. *Courtesy of Doug and Pat Wengel.* $100-$250.

"Donald at Beach" sand pail. Walt Disney Productions. Early 1940s. 5" tall. *Courtesy of Doug and Pat Wengel.* $175-$350.

Two views "Donald Tug of War" sand pail. Walt Disney Productions. December, 1939. Sometimes referred to as one of the "switchover" pails, this pail was produced right after Disney changed from Walt Disney Enterprises to Walt Disney Productions in December 1939. 4" tall. *Courtesy of Doug and Pat Wengel.* $225-$400.

"Muscle Beach Donald Duck" sand pail. Walt Disney Productions. Another "switch-over" piece. 2 3/4" tall. *Courtesy of Doug and Pat Wengel.* $125-$250.

"Donald and Nephew Musicians" sand pail. Walt Disney Productions. *Courtesy of Doug and Pat Wengel.* $125-$250.

Sand Sets

NO. 12 SAND SIEVE SET
"SEASHORE KIDS" design. Size: 6½ x 1 5/8". Contains fish, spoon & round shell. With screen bottom. 6 dozen to case. 4 lbs. per dozen.

NO. 55 SAND SIEVE SET
"SEASHORE KIDS" design. Size: 6½ x 1 5/8". Contains fish, small shovel & round shell. With screen bottom. 6 dozen to case. 4 lbs. per dozen.

NO. 13 SAND CARD SET
On 7 x 10" card containing frog, fish, spoon, & sieve. 3 dozen to case. 4 lbs. per dozen.

NO. 14 SAND CARD SET
On 7 x 10" card containing pail, spoon, fish & perforated sieve. 3 dozen to case. 4½ lbs. per dozen.

Sand Hoist   Water Pump

NO. 20 SAND HOIST
Silver, red & blue colors on parts. 8" long by 11" high. Cart automatically dumps when full. 2 dozen to case. 13 lbs. per dozen.

NO. 21 KIDDIE PLAY PUMP
"FIRE KIDS" design. 10" high. Tub 2½ x 6". Small water bucket included. 2 dozen to case. 12 lbs. per dozen.

1942 catalogue page

"Donald at Sea on Raft" sand pail. Walt Disney Productions. 1940s. 3 1/4" tall. *Courtesy of Doug and Pat Wengel.* $125-$225.

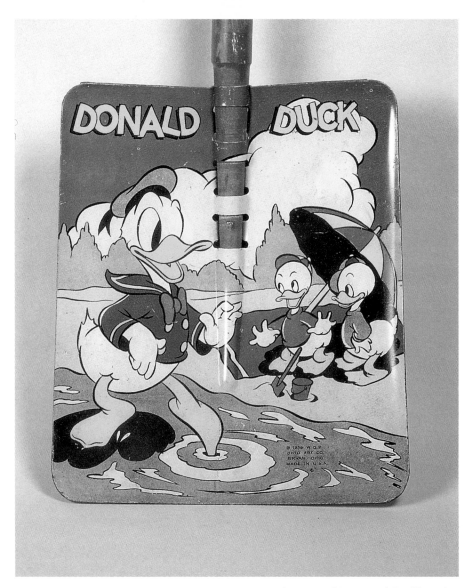

Donald Duck sand shovel. Walt Disney Productions. *Courtesy of Doug and Pat Wengel.* $100-$145.

Donald Duck Sand Sifter. Walt Disney Productions. *Courtesy of Jim Gilcher.* $85-$125.

"Kids at Beach" sand sifter and sand mold set. 1940s. $65-$95.

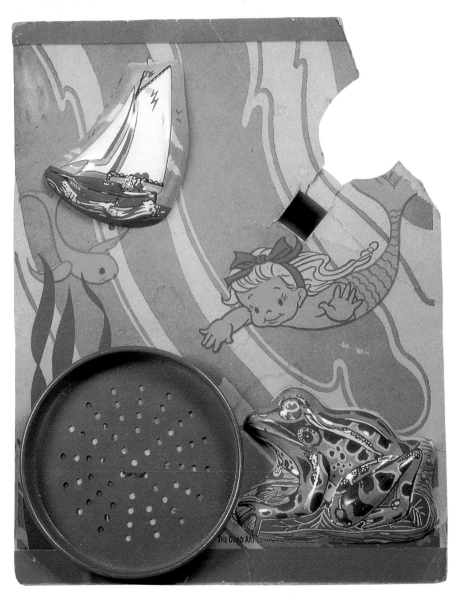

Sand mold set. Mint-on-card: $75.

Sand Mold set. No. 13. 1942, 1943, 1949, 1950. Mint-on-card: $75.

Elephant water pump.
1940s. $65-$100.

"Ducky" water
pump. 1940s.
$65-$100.

"Fireman" water pump with box. Late 1940s through 1950s. This pump came with little bucket. *Courtesy of Jim Gilcher.* $65-$125.

Watering Cans

NO. 8 WATERING CAN
"DONALD DUCK" design. Size: 7½ x 5 ¾". 6 dozen to case. 4 lbs. per dozen.

NO. 7 WATERING CAN
"MODERN FARMER" design. Size: 5½ x 3". 12 dozen to case. 1⅗ lbs. to dozen.

NO. 9 WATERING CAN
"ROMEO & JULIET" designs. Size: 7½ x 5 ¾". 6 dozen to case. 4 lbs. per dozen.

NO. 11 LONG SPOUT CAN
"FLORAL" design. Size 13½ x 5¼". 1 dozen to case. 7 lbs. per dozen.

NO. 10 WATERING CAN
"BO PEEP" design. Size: 11 x 8½". 2 dozen to case. 12 lbs. per dozen.

Shovels

NO. 30 SAND SHOVEL
"DONALD DUCK" design. 23" handle, 6½ x 7½" blade. 6 dozen to case. 6⅗ lbs. per dozen.

NO. 29 SAND SHOVEL
"DONALD DUCK" design. 15" handle, 4 x 5" blade. 6 dozen to case. 3½ lbs. per dozen.

NO. 31 SCOOP SHOVEL
All red colors. 9" long overall, blade 3 x 3¾". 12 dozen to case. 3 lbs. per dozen.

NO. 32 SCOOP SHOVEL
All red colors. 14" long overall, blade 4½ x 5½". 6 dozen to case. 6⅗ lbs. per dozen.

1942 catalogue page.

Top right & bottom right: "Donald Duck" sprinkling can. No. 29. Walt Disney Enterprises. 1940. A "cross-over" piece. This tiny, cute can has a before-and-after scene of Donald tripping over a brick and losing his famous temper. $100-$135.

"Donald on Train" sprinkling can. Walt Disney Productions. 1940s. *Courtesy of Jim Gilcher*. $150-$235.

Watering can. 1940s. Elaine Ends Hileman.
Courtesy of Ohio Art Company. $50-$75.

Two watering cans from the 1940s. The smaller one is by Elaine Ends Hileman. *Courtesy of Jim Gilcher.* $50-$75.

"Mary Mary Quite Contrary" watering can. Late 1930s through early 1940s. Fern Bisel Peat. Large size. $125-$175.

89

Above: "Child, Ball, and Animal" drum. Mid-1940s. Fern Bisel Peat. *Courtesy of Ohio Art Company.* $45-$75.

Above right: "Marching" drum. Early 1940s. Fern Bisel Peat. *Courtesy of Ohio Art Company* $45-$75.

Bottom right: "Boys and Girls Parade" drum. Early 1940s. HK Beatrice Benjamin. $35-$65.

"Bunny" top. 1940s.
Hileman. Small. *Courtesy
of Jim Gilcher.* $45-$70.

"Children at Play" top.
1940s. $35-$75.

Three 1940s tops. "Dutch girl,"
"Zebras," and "Clowns." *Courtesy
of Jim Gilcher.* $40-$60.

"Children with Donkeys" top.
1940s. *Courtesy of Jim Gilcher.*
$35-$65.

Music box with original box.
1940s. *Courtesy of Jim Gilcher.*
$45-$75.

Drum Bank. 1941. Fern Bisel Peat. This band was produced for several years.
$35-$60.

"Donald Duck, Housekeeper" carpet sweeper. Walt Disney Productions. 1941-43. $150-$225.

"Musical Sweeper." Late 1940s. *Courtesy of the Ohio Art Company.* $45-$70.

Wheelbarrow. Early 1940s. *Courtesy of Jim Gilcher.* $35-$50.

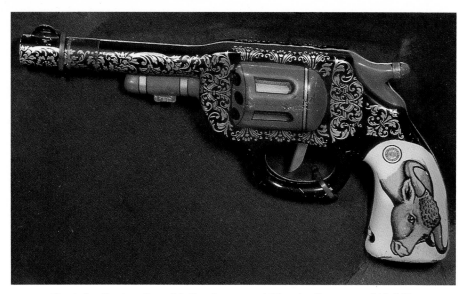

Tin-Litho Cowboy Gun. 1940s. *Courtesy of Ohio Art Company.* $45-$80.

Christmas specialty item. 1949 through the 1950s. $50-$135.

Easter Bunny Cart. 1940s. *Courtesy of Jim Gilcher.* $50-$85.

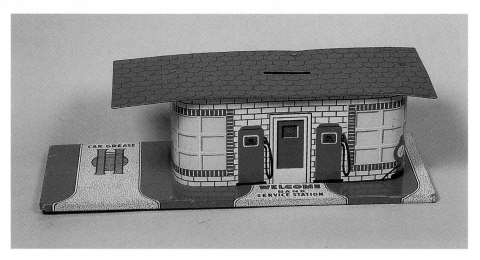

Service Station Bank. 1949-51. *Courtesy of Ohio Art Company.* $30-$50.

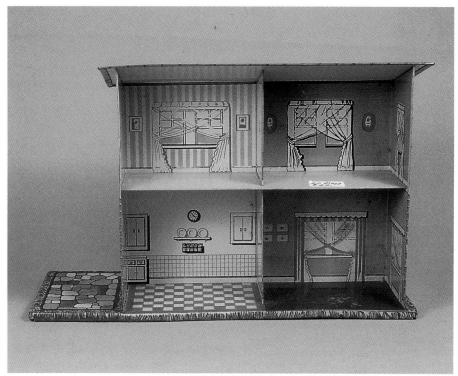

Exterior and interior of tiny doll's house. 1949-52. $40-$75.

Metal Beverage Coasters

DESIGN 1   DESIGN 2   DESIGN 3

DESIGN 4   DESIGN 5   DESIGN 6

NO. 28 SERIES COASTERS

Liquor Proof Laquer ★ 3¼ Inch Diameter ★ Indented for Stacking

In revamping the number 28 line of "LITHOGRAPHED METAL BEVERAGE COASTERS" many improved changes have been made. The line now contains six very "snappy" designs that are sure to fill almost any requirement of the customer at the counter. The indentation around the edge of the coaster allows the item to be stacked without side slip. The small, inconspicuous bottom indentations prevent the glass from sticking to the coaster when a little liquid is spilled.

Coasters are wrapped one dozen to a bundle, packed in a 24 dozen shipping case. If the standard assortment is wanted, we will ship 4–1 dozen bundles of each of the above six designs. If you do not want the standard assortment but would rather make up your own, do so by ordering by the number shown on illustrations above. Designs must be ordered in even dozens however and the minimum shipment must be 24 dozen pieces. Please keep all total orders in multiples of 24 dozen cases as these are tightly packed to prevent any damage. **Case of 24 dozen weighs 15 lbs.**

1942 catalogue page showing metal beverage coasters.

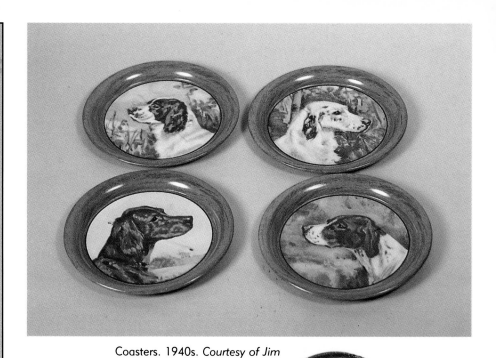

Coasters. 1940s. *Courtesy of Jim Gilcher.* $15-$40.

Popcorn Snack Set. 1940s. *Courtesy of Jim Gilcher.* $50-$75.

"Butt Bucket." Late 1940s-50s. Don Dean. This totally politically incorrect item has fabulous and very witty lithography including little bluebirds smoking stogies! $65-$100.

Drum Lamp. Produced between 1945 and 1959. *Courtesy of Jim Gilcher.* $50-$145.

FRONTIER DAYS PRAIRIE WAGON

CALIFORNIA OR BUST!

WELLS FARGO & CO.

CHAPTER V: THE 1950S

In 1951, Ohio Art was awarded the United States Navy contract for the manufacture of rocket motor parts for use in the Korean War. In 1953, the company introduced its first internally manufactured plastic toys, including a Farm Animal Set, a Farmyard Set, several tea sets, and a sand sieve set. The company continued to produce tin-litho tea sets and sand toys as well as several new types of toys including wind-up toys and toy trucks.

The trucks were the result of Ohio Art's purchase of the Dunwall Company whose inventory consisted of a line of toy trucks. After the purchase, Ohio Art sold the trucks under the name "Buckeye." Ohio Art never tooled the "Buckeye" trucks; they simply marketed the inventory.

Buckeye Truck. *Courtesy of Jim Gilcher.* $350-$500.

FINEST MADE...FASTEST SOLD

BUCKEYE TRUCKS

SnowCrop FROZEN FOODS

The OHIO ART Company

Bryan, Ohio

U. S. A.

1959 catalogue page advertising Buckeye trucks.

Buckeye Truck. *Courtesy of John Walters.* $350-$500.

Little Chef

- Sturdy Steel construction
- Rivited and spot welded
- Scientifically scaled for small-fry pleasure
- Sleek white appliance enamel finish
- It cooks! It bakes! It's safe!

UL

NO. 162 PLAY STOVE & UTENSILS

Lots of play hours in this attractive and "jumbo" size play stove. Done in gleaming white enamel and has 4 gas burners on top-very realistic looking. Sturdy metal construction. Oven door opens. Red plastic turning knobs. 3 piece high quality aluminum utensils as shown. Shiny control panel with simulated clock. Size 13 3/8 x 12¾ x 8". Each in corr. case. ½ doz. ctn. 38 lbs.

High Quality - Low Price!

NO. 163 ELECTRIC RANGE AND UTENSILS

A real electric model stove. Large Alloyed metal cooking surface and oven to keep food warm. White baked enamel finish-enclosed heating element. Five piece set of high quality aluminum cooking utensils as shown. Completely Underwriter's Approved. 110 Volt. 11½ x 10 1/8 x 7". Each in carton. ½ doz. ctn. weighs 32 lbs.

Really Deluxe!

NO. 164 DELUXE "LITTLE CHEF" ELECTRIC RANGE

Comes complete with 6 pc. set high quality aluminum utensils as shown. This is the finest range combination ever made. Gleaming white baked enamel-chrome trim, plastic "push-buttons" for added play. Separate elements for top burner and oven. This stove really bakes most of the food set recipes now being sold thru toy departments. Window in oven door. Both oven doors open. Underwriter's Approved. 13 3/8 x 12 3/4 x [...]. Each in carton. One half dozen master, 56 lbs. [...] ch in carton weighs 9 lbs.

11

1956 catalogue page showing "Little Chef" items.

The Quality Line of Play & Electric Stoves & Utensils

● WONDERFUL VALUES ● HOURS of PLAY FUN

GLEAMING STAINLESS STEEL IS CONTRASTED WITH JET BLACK BAKELITE HANDLES AND COPPER BOTTOMS

Set Contains "TRIG" Tea Kettle - Tops In The World!

NO. 109 "LITTLE CHEF" MINIATURE WARE SET

This beautiful 8 piece cooking set is a perfect miniature of full size utensils. The jewel-like precision in scale and in manufacture makes this set the top value on the market. Every detail of mother's kitchen ware has been duplicated. Contains: Frying Pan, Cooking Pan, Lid, Spatula, Cake Knife, Stirring Spoon and Utensil rack for wall mounting. Famous "Trig" Whistling Tea Kettle. Display box of shiny silver foil can be hung on wall or used as floor cabinet. ½ doz. ctn. 13 lbs.

NO. 109

The No. 109 is entirely repackaged this year and the new unit is of shiny silver foil-made well to hang on wall as cabinet.

NO. 133 MECHANICAL SPARKING STOVE

Complete modern range with 9 piece high impact utensil set. Mechanical motor is wound in back and started by front knob. Two ruby burners shoot sparks under pans. Oven door opens and has ruby front. Utensils held in shiny top rack. 8½ x 8 x 4½". Each in printed box. 1 doz. ctn. 10 lbs.

NO. 130 PLAY RANGE

Beautiful design and litho. Oven door opens. Burners are raised. 9 piece high impact utensil set. 5 1/8" high, 8" wide, 4½" deep. Each in printed box. 1 doz. ctn. 9 lbs.

10

Cookware. 1956.

In 1955 Ohio Art purchased "Tacoma Metal Products Company" of Tacoma, Washington. The Company manufactured toy electric stoves that actually baked and Ohio Art marketed the stoves as part of its "Little Chef" line.

Stove. Late 1950s. *Courtesy of Ohio Art Company.* $75-$125.

The popularity of cowboys is reflected in much of the Ohio Art lithography of the 1950s. Ohio Art used Roy Rogers' image on several of its new toys including the Cowboy lantern advertised on the cover of its 1956 catalogue.

Roy Rogers lantern. 1956. $90-$150.

Lithographed METAL and RUBBER-LIKE VINYL TOYS

"The Quality Line Since 1909"

Fall & Winter 1956

This new "Roy Rogers" Ranch Lantern described on Page 3.

The Ohio Art Co., Bryan, Ohio

1956 catalogue page showing Roy Rogers lantern.

There are so many other fifties toys, from dishes to wind-ups. Toys in the fifties were still charming and the lithography was still of exceptional quality. See for yourselves!

More little cowboy favorites.

1956 catalogue showing horseshoe set.

Right: Cowboy horseshoes set. 1955-57. *Courtesy of John Walters.* $45-$75.

102

1957 catalogue page showing Chuckwagon.

Below: Chuckwagon. *Courtesy of Ohio Art Company.* $400-$500.

NO. 70 PLAID FLOWERS

Twenty-one piece Tea Set. All pieces securely attached to filler. Bright litho colors — sturdy metal parts. Consists of 8x10" tray, 4 cups, 4 saucers, 4-4" plates, 4 butter plates, footed sugar bowl, tall creamer, tea pot and lid.
Packed 1 dozen to case.
22 lbs. per dozen.

NO. 71 "CIRCUS"

Twenty-one piece Tea Set. All pieces securely attached to filler. Bright litho colors — sturdy metal parts. Consists of 8x10" tray, 4 cups, 4 saucers, 4-4" plates, 4 butter plates, footed sugar bowl, tall creamer, tea pot and lid.
Packed 1 dozen to case.
22 lbs. per dozen.

NO. 72 "BLUE WILLOW"

Twenty-one piece Tea Set. All pieces securely attached to filler. Bright litho colors — sturdy metal parts. Consists of 8x10" tray, 4 cups, 4 saucers, 4-4" plates, 4 butter plates, footed sugar bowl, tall creamer, tea pot and lid.
Packed 1 dozen to case.
22 lbs. per dozen.

NEW! WON'T BREAK! POLY PLASTIC! LIFE SIZE PARTS!

NEW

NO. 193 PASTRY-CANISTER SET

16 pc. set contains 3 metal canisters with lids (Largest is 3x3" — others in proportion) Pastry board, 2 pie pans, 3 metal bowls, rolling pin, cookie sheet, shell mold and large poly mixing spoon.
1 dozen to case.
16 lbs. per dozen.

NO. 159 POLY TEA SET

Twenty piece set of beautifully colored soft, flexible poly plastic. All life size pieces. 3 cups, 3 saucers, 3 plates, 3 butter plates, 3 glasses, 3 spoons, tea pot and lid.
½ dozen to case.
20 lbs per dozen.

NO. 160 POLY TEA SET

Twenty-six pc. set of beautifully colored soft, flexible poly plastic. All life size. 4 cups, 4 saucers, 4 plates, 4 butter plates, 4 glasses, 4 spoons, tea pot and lid.
½ dozen to case.
23 lbs. per dozen.

PLEASE SEE ADDITIONAL TEA SETS NEXT PAGE

1955 catalogue page showing "Red Plaid," "Circus" tea set, "Blue Willow" tea set, "Cherry" canister set, and "Fiesta Ware."

"Red Plaid" tea set. Mid-1950s. Four place setting with large coffee pot and a pitcher with lid. Four piece place setting with large coffee pot and pitcher with lid: $125-$225.

"Circus" tea set. Early through mid-1950s. Complete thirty-one-piece set: $150-$225.

"Blue Willow" tea set. Early through mid-1950s. *Courtesy of Jim Gilcher.* Four-piece place setting: $90-$150.

NO. 61 TEA SET

Seven piece set. Four color litho design. 4" tray, 2 cups, 2 saucers, tea pot and lid. 6 doz. ctn. 23 lbs.

NO. 63 TEA SET

Eight piece set. Four color litho design. 5" x 7" tray, 2 cups, 2 saucers, 2 plates, tea pot. 2 doz. ctn. 19 lbs.

NO. 64 TEA SET

Fourteen piece set, four color litho. 5 x 7" tray, 4 cups 4 saucers, 4 plates, tea pot. 2 doz. ctn. 26 lbs.

NO. 65 TEA SET

Fourteen piece set. Four color litho design. 5" x 7" tray, 4 cups, 4 saucers, 4 plates, tea pot. 2 doz. ctn. 30 lbs.

All Metal Tea Sets Have Vinyl-Non Rust Varnish

NO. 79 TEN PIECE REFRIGERATOR SET

Sturdy, bright 3x5" metal litho tray, 4 Ruby Plastic Goblets, 4 Ruby Sherbets, 1 large blue water pitcher. 2 doz. ctn. 15 lbs.

Soft, Flexible Plastic Sets
Terrific New Packaging!

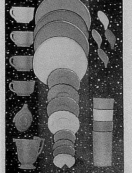

NO. 159 POLY SET

13 Pc. set of soft, poly plastic in life-size pcs. Pot, 2 each Tumblers, Cups, Saucers, Plates, Butter plates, Spoons. 1 doz. ctn. 14 lbs.

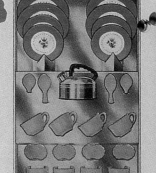

NO. 160 POLY SET

26 Pc. set of soft, poly plastic in life-size pcs. Pot, Lid, 4 each Tumblers, Cups, Saucers, Plates, Butter Plates, Spoons. ½ doz. ctn. 12 lbs.

NO. 161 POLY & METAL COMBINATION

Novel combination of poly and metal. One "TRIG" Aluminum Tea Kettle w/copper bot. trigger action. 4 each POLY parts Tumbler, Butter Plates, Spoons. 4 each METAL Saucers and Plates. Foil Package. ½ doz. ctn. 10 lbs.

4

1956 catalogue page showing more tea sets.

"Boy and Girl in Garden" tea set. 1956. *Courtesy of Jim Gilcher.* Fifteen-piece set: $65-$150.

"She Loves Me, She Loves Me Not," tea set. This set was issued in 1949 with straight-sided cups and then again in 1956 with rounded cups. *Courtesy of Jim Gilcher.* Six-piece set: $50-$125.

"Apple and Pear" tea set. Mid-1950s. *Courtesy of Jim Gilcher.* Seven-piece set: $40-$75.

"The Wedding" tea set. 1950s. This set remains a mystery. It seems clear that it is representative of particular story, but despite our efforts, we have not determined which story it could be. *Courtesy of Jim Gilcher.* Thirty-one-piece set: $150-$250.

"Girl with Dolly" tea set. Mid-1950s. *Courtesy of Jim Gilcher.* Eleven-piece set: $125-$160.

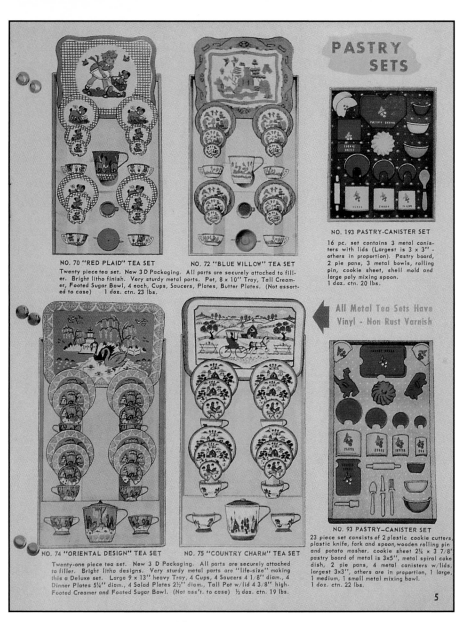

NO. 70 "RED PLAID" TEA SET
Twenty piece tea set. New 3 D Packaging. All parts are securely attached to filler. Bright litho finish. Very sturdy metal parts. Pot, 8 x 10" Tray, Tall Creamer, Footed Sugar Bowl, 4 each, Cups, Saucers, Plates, Butter Plates. (Not assorted to case) 1 doz. ctn. 23 lbs.

NO. 72 "BLUE WILLOW" TEA SET

NO. 193 PASTRY-CANISTER SET
16 pc. set contains 3 metal canisters with lids (Largest is 3 x 3" - others in proportion). Pastry board, 2 pie pans, 3 metal bowls, rolling pin, cookie sheet, shell mold and large poly mixing spoon. 1 doz. ctn. 20 lbs.

All Metal Tea Sets Have Vinyl - Non Rust Varnish ◀

NO. 74 "ORIENTAL DESIGN" TEA SET

NO. 75 "COUNTRY CHARM" TEA SET
Twenty-one piece tea set. New 3 D Packaging. All parts are securely attached to filler. Bright litho designs. Very sturdy metal parts are "life-size" making this a Deluxe set. Large 9 x 13" heavy Tray, 4 Cups, 4 Saucers 4 1/8" diam., 4 Dinner Plates 5¼" diam., 4 Salad Plates 2½" diam., Tall Pot w/lid 4 3/8" high, Footed Creamer and Footed Sugar Bowl. (Not ass't. to case) ½ doz. ctn. 19 lbs.

NO. 93 PASTRY—CANISTER SET
23 piece set consists of 2 plastic cookie cutters, plastic knife, fork and spoon, wooden rolling pin and potato masher. cookie sheet 2¾ x 3 7/8" pastry board of metal is 3x5", metal spiral cake dish, 2 pie pans, 4 metal canisters w/lids, largest 3x3", others in proportion, 1 large, 1 medium, 1 small metal mixing bowl. 1 doz. ctn. 22 lbs.

5

1956 catalogue page.

"Pink Swans" tea set. 1956. This wonderfully campy set has a marvelous motif and fabulous color. Swans and the pink and black color scheme were very popular in the 1950s. Twenty-one piece set: $125-$175.

"Cherries" canister set. No. 193. 1956. *Courtesy of Shirley Cox.* $150-$300.

"Red and Green Floral" tea set. This set was produced from the 1940s through the early 1950s. Thirty-one-piece set: $150-$250.

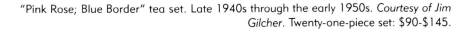

"Dutch Winter Wonderland" tea set. 1950s. Fifteen-piece set: $140-$200.

"Pink Rose; Blue Border" tea set. Late 1940s through the early 1950s. *Courtesy of Jim Gilcher.* Twenty-one-piece set: $90-$145.

"Cinderella" tea set. Circa 1952. Eleven-piece set: $125-$175.

"Lady and Gentleman" tea set. 1959-61. Twenty-one-piece set: $90-$150.

"Coffee Time" tea set. 1959-62. Nine-piece set: $65-$110.

"Noah's Ark" sand pail. 1956. This cute pail has a hole in the top of the handle. The shovel fits through the hole to make a sail. *Courtesy of Jim Gilcher.* $100-$150.

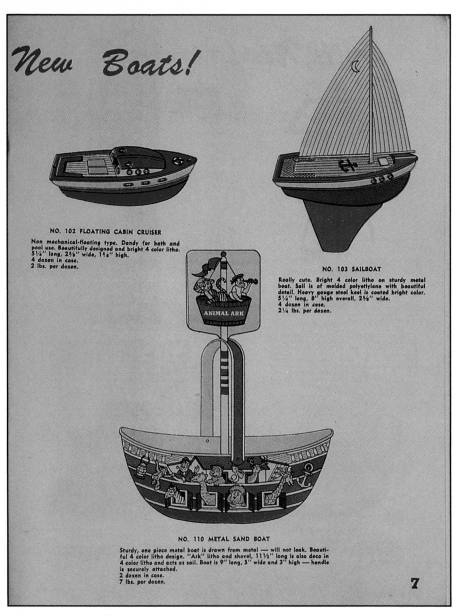

New Boats!

NO. 102 FLOATING CABIN CRUISER

Non mechanical-floating type. Dandy for bath and pool use. Beautifully designed and bright 4 color litho. 5¼" long, 2½" wide, 1⅝" high.
4 dozen in case.
2 lbs. per dozen.

NO. 103 SAILBOAT

Really cute. Bright 4 color litho on sturdy metal boat. Sail is of molded polyetlylene with beautiful detail. Heavy gauge steel keel is coated bright color. 5¼" long, 8" high overall, 2⅝" wide.
4 dozen in case.
2¼ lbs. per dozen.

ANIMAL ARK

NO. 110 METAL SAND BOAT

Sturdy, one piece metal boat is drawn from metal — will not leak. Beautiful 4 color litho design. "Ark" litho and shovel, 11½" long is also deco in 4 color litho and acts as sail. Boat is 9" long, 5" wide and 3" high — handle is securely attached.
2 dozen in case.
7 lbs. per dozen.

7

1955 catalogue page showing boats and sand pails.

"Davy Crockett" sand pail. Walt Disney Productions. 1955. Large size. *Courtesy of Doug and Pat Wengel.* $110-$160.

"Children at the Beach" sand pail. 1950. Medium size. $75-$125.

"Cowboy" sand pail. Small size. 1953-55. $45-$90.

"Children and Puppies at the Beach" sand pail. 1950. Small size. $50-$100.

"Gardner and Red Brick Wall" sand pail. 1950s. Medium size. $60-$85.

"Cowboys and Indians" sand pail. Early 1950s. Medium size. *Courtesy of Claudette Job.* $65-$125.

"Clowns" sand pail. 1953-55. Small size. $45-$65.

"Huckleberry Hound" sand pail. 1959-61. Large size. *Courtesy of Jim Gilcher.* $50-$100.

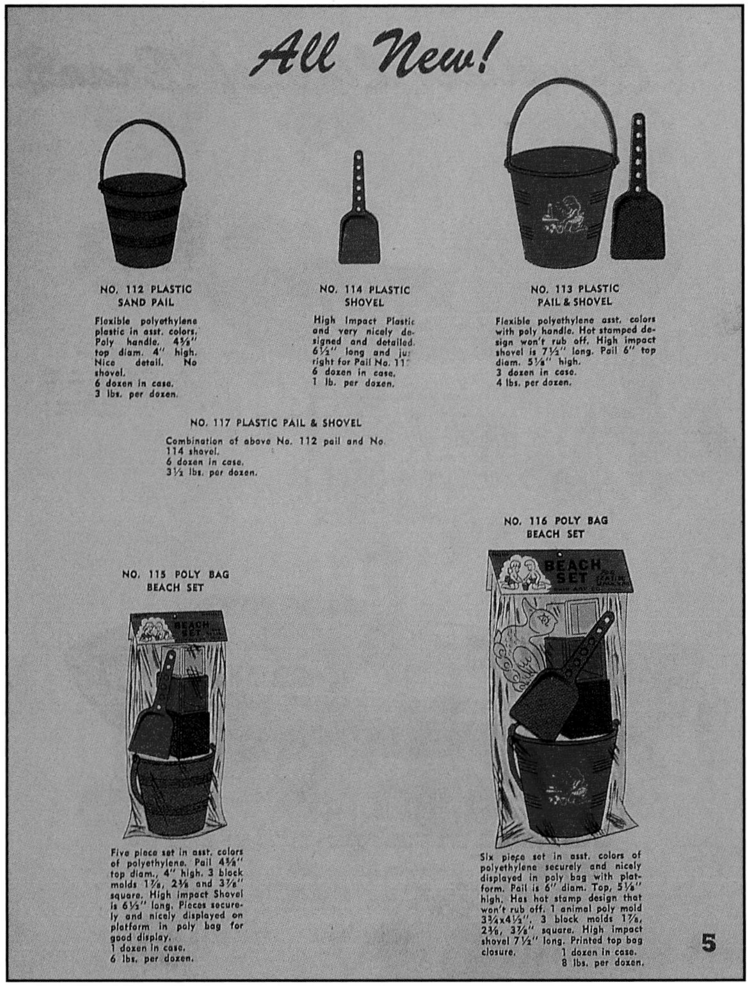

All New!

NO. 112 PLASTIC SAND PAIL
Flexible polyethylene plastic in asst. colors. Poly handle. 4⅝" top diam. 4" high. Nice detail. No shovel.
6 dozen in case.
3 lbs. per dozen.

NO. 114 PLASTIC SHOVEL
High Impact Plastic and very nicely designed and detailed. 6½" long and ju[illegible] right for Pail No. 11[illegible]
6 dozen in case.
1 lb. per dozen.

NO. 113 PLASTIC PAIL & SHOVEL
Flexible polyethylene asst. colors with poly handle. Hot stamped design won't rub off. High impact shovel is 7½" long. Pail 6" top diam. 5⅛" high.
3 dozen in case.
4 lbs. per dozen.

NO. 117 PLASTIC PAIL & SHOVEL
Combination of above No. 112 pail and No. 114 shovel.
6 dozen in case.
3½ lbs. per dozen.

NO. 115 POLY BAG BEACH SET
Five piece set in asst. colors of polyethylene. Pail 4⅝" top diam., 4" high. 3 block molds 1⅞, 2⅜ and 3⅞" square. High impact Shovel is 6½" long. Pieces securely and nicely displayed on platform in poly bag for good display.
1 dozen in case.
6 lbs. per dozen.

NO. 116 POLY BAG BEACH SET
Six piece set in asst. colors of polyethylene securely and nicely displayed in poly bag with platform. Pail is 6" diam. Top, 5⅛" high. Has hot stamp design that won't rub off. 1 animal poly mold 3¾x4½". 3 block molds 1⅞, 2⅜, 3⅞" square. High impact shovel 7½" long. Printed top bag closure.
1 dozen in case.
8 lbs. per dozen.

5

1955 catalogue page showing plastic sandpails.

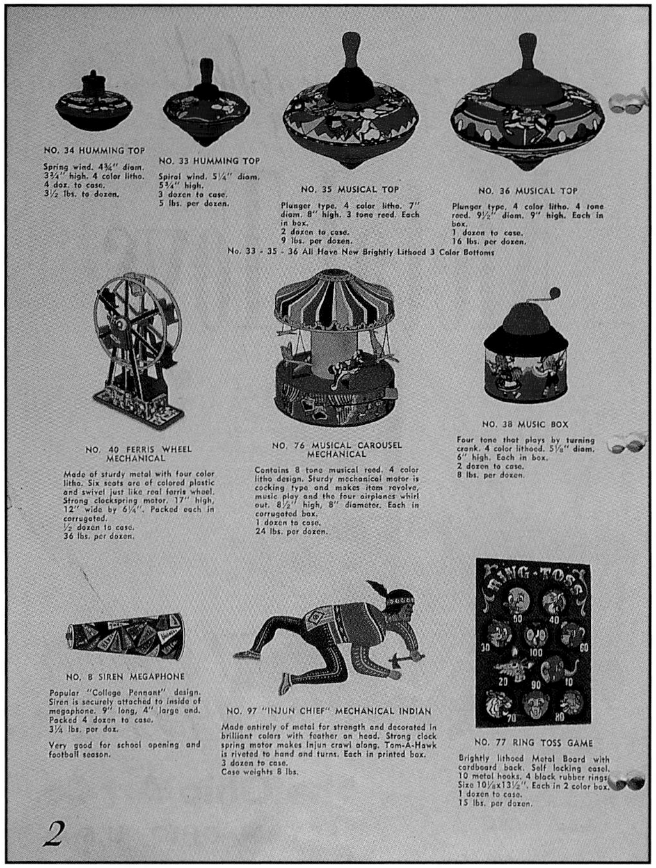

NO. 34 HUMMING TOP
Spring wind. 4¾" diam. 3¾" high. 4 color litho.
4 doz. to case.
3½ lbs. to dozen.

NO. 33 HUMMING TOP
Spiral wind. 5¼" diam. 5¾" high.
3 dozen to case.
5 lbs. per dozen.

NO. 35 MUSICAL TOP
Plunger type. 4 color litho. 7" diam. 8" high. 3 tone reed. Each in box.
2 dozen to case.
9 lbs. per dozen.

NO. 36 MUSICAL TOP
Plunger type. 4 color litho. 4 tone reed. 9½" diam. 9" high. Each in box.
1 dozen to case.
16 lbs. per dozen.

No. 33 - 35 - 36 All Have New Brightly Lithoed 3 Color Bottoms

NO. 40 FERRIS WHEEL MECHANICAL
Made of sturdy metal with four color litho. Six seats are of colored plastic and swivel just like real ferris wheel. Strong clockspring motor. 17" high, 12" wide by 6¼". Packed each in corrugated.
½ dozen to case.
36 lbs. per dozen.

NO. 76 MUSICAL CAROUSEL MECHANICAL
Contains 8 tone musical reed. 4 color litho design. Sturdy mechanical motor is cocking type and makes item revolve, music play and the four airplanes whirl out. 8½" high, 8" diameter. Each in corrugated box.
1 dozen to case.
24 lbs. per dozen.

NO. 38 MUSIC BOX
Four tone that plays by turning crank. 4 color lithoed. 5⅛" diam. 6" high. Each in box.
2 dozen to case.
8 lbs. per dozen.

NO. 8 SIREN MEGAPHONE
Popular "College Pennant" design. Siren is securely attached to inside of megaphone. 9" long, 4" large end. Packed 4 dozen to case.
3¼ lbs. per doz.

Very good for school opening and football season.

NO. 97 "INJUN CHIEF" MECHANICAL INDIAN
Made entirely of metal for strength and decorated in brilliant colors with feather on head. Strong clock spring motor makes Injun crawl along. Tom-A-Hawk is riveted to hand and turns. Each in printed box.
3 dozen to case.
Case weights 8 lbs.

NO. 77 RING TOSS GAME
Brightly lithoed Metal Board with cardboard back. Self locking easel. 10 metal hooks. 4 black rubber rings. Size 10⅛x13½". Each in 2 color box.
1 dozen to case.
15 lbs. per dozen.

2

1953 catalogue page showing tops and wind-up toys.

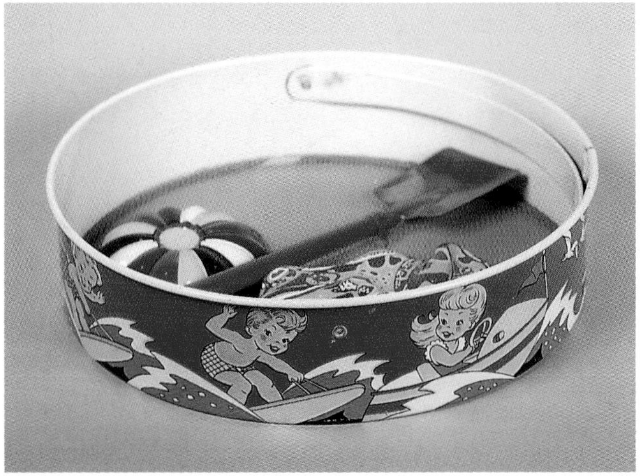

"Water Skiers" sand sieve. $50-$85.

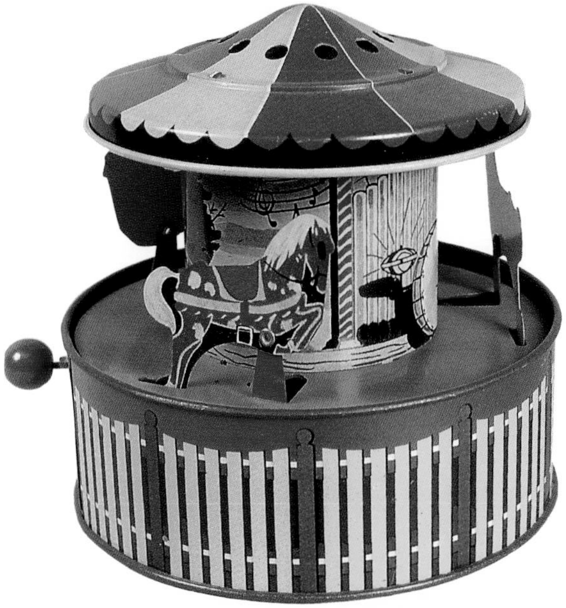

Merry-Go-Round. *Courtesy of Ohio Art Company*. $150-$250.

Ferris Wheel. *Courtesy of Ohio Art Company*. $225-$350.

Whirlybird Target. 1953-54. Large. *Courtesy of Jim Gilcher.* $60-$100.

Doll Buggy. No. 57. 1954. 6 3/4" long. *Courtesy of Jim Gilcher*. $35-$75.

Megaphone. This toy had a siren whistle built into the top. *Courtesy of Jim Gilcher*. $25-$50.

Battery operated lantern. $50-$75.

"GI Joe" windup. Courtesy of John Walters. $50-$150.

Target game. Late 1950s-60s. The same dies were used for shooting galleries of different designs. Courtesy of Jim Gilcher. $60-$100.

Wind-up turtle. Courtesy of Jim Gilcher. $35-$50.

Airport with long-term action. *Courtesy of Jim Gilcher.* $200-$350.

Ski plane. *Courtesy of Jim Gilcher.* $65-$115.

Musical wind up dog house. *Courtesy of Jim Gilcher.* $50-$75.

SOFT RUBBER-LIKE VINYL TOYS

NO. 190 SHEEP	NO. 191 PIG	NO. 192 CALF	NO. 194 COLT	NO. 186 COW	NO. 187 HORSE
2" long.	2¼" long.	2" long.	2½" long.	3 5/8" long.	3 5/8" long.
6 doz. ctn. 2¼ lbs.	6 doz. ctn. 3½ lbs.	6 doz. ctn. 2 lbs.	6 doz. ctn. 3 lbs.	2½" high	3 5/8" high
				4 doz. ctn. 3½ lbs.	4 doz. ctn. 4 lbs.

ALL ANIMALS IN ASSORTED VINYL COLORS

NO. 189 33 PC. VINYL ANIMAL ASST.

Rubber-like vinyl animals asstd: 2 cows, 1 horse, 2 sheep, 2 colts, 2 pigs, 2 roosters, 2 duck families, 3 calves, 1 turkey, 4 piglets, 4 lambs, 8 hens. Each set attractively packaged. 1 doz. ctn. 11 lbs.

NO. 188 20 PC. VINYL ANIMAL ASST.

Rubber-like vinyl animals asstd: 10 colts, calves, pigs, sheep. 10 hens, piglets, ducks, turkey, rooster. Ea. set attractively packaged. 1 doz. ctn. 7 lbs.

NO. 195 BARN & ANIMALS

Bright litho barn with Cupola and Silo and top. Barn 8½ x 5¾ x 4¼". 9 soft plastic animals & 4 pieces fence. Each in box. 1 doz. ctn. 14 lbs.

NO. 88 BARN & ANIMALS

Beautifully designed new Animals are of soft rubber-like vinyl-nicely detailed. Plastic Tractor & Wagon. 12 sections fence. Same set of Animals as in above No. 189 33 Piece set. Barn beautifully lithoed inside and outside. Silo with top hooks to barn-sliding door in barn front. Barn is 19 x 8½ wide x 11" high. Each in attractive counter display box with instructions. ½ doz. ctn. 27 lbs.

9

Above: 1959 catalogue showing farm sets.

Top left: Box for 1956-57 farm set. *Courtesy of Jim Gilcher.*

Left: Farm set with box. *Courtesy of Jim Gilcher.* $60-$100.

Doll Strollers

Top Value Sales Producers

NO. 132 DOLL STROLLER

Brand new and beautifully designed and lithoed. Sturdy metal and pastel color trimmed. Black vinyl tires are non marking and 3¾" diam. Sturdy metal axels and rounded, smooth hubs. Handle 20" from floor. Back of Seat 15¼" from floor. Seat is 7 x 8" with rolled edges. Front of seat is heavy wire with 4 colored wooden balls. 20" high, 10¼" wide and 15½" long overall. Handle folds down and body drops into wheels for compact shipment–2 bolts hold all together rigidly. Each in corr. box. ½ doz. ctn. 20 lbs.

NO. 139 STROLLER

Petite size just right for new popular 8" dolls. Sturdy metal brightly lithoed four colors. 5¼" high, 4" wide. 16" yellow handle with red knob. Wheels are 2¼" diam. Very good candy pack item. 2 doz. ctn. 8 lbs.

NO. 125 DOLL STROLLER

Top dollar producer in our line last year. Tremendous value. Done in pastel colors of trim-bright litho design on inside and outside body and wheels. Sturdy metal. Plastic strap and bucket to hold up to 18" doll. 11" high, 8½" wide. 5" diam. wheels. Colored handle and ball of wood. Handle 18" from floor. No assembly other than screw in handle. ½ doz. ctn. 9 lbs.

6

1956 catalogue showing strollers.

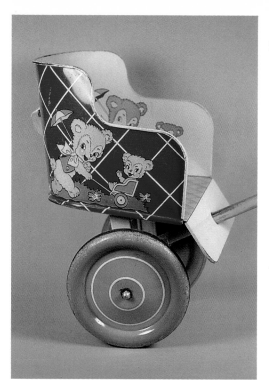

1950s stroller. Early to mid-1950s. $45-$65.

Stroller. Produced 1959-61.
Courtesy of Jim Gilcher. $50-$90.

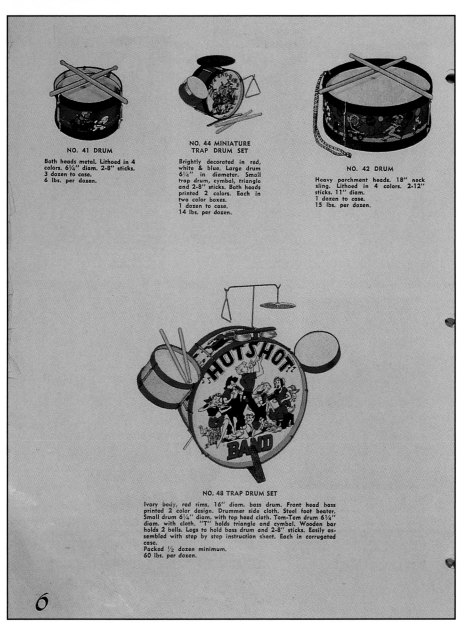

NO. 41 DRUM
Both heads metal. Lithoed in 4 colors. 6¼" diam. 2-8" sticks. 3 dozen to case. 6 lbs. per dozen.

NO. 44 MINIATURE TRAP DRUM SET
Brightly decorated in red, white & blue. Large drum 6¼" in diameter. Small trap drum, cymbal, triangle and 2-8" sticks. Both heads printed 2 colors. Each in two color boxes. 1 dozen to case. 14 lbs. per dozen.

NO. 42 DRUM
Heavy parchment heads. 18" neck sling. Lithoed in 4 colors. 2-12" sticks. 11" diam. 1 dozen to case. 15 lbs. per dozen.

NO. 48 TRAP DRUM SET
Ivory body, red rims, 16" diam. bass drum. Front head bass printed 2 color design. Drummer side cloth. Steel foot beater. Small drum 6¼" diam. with top head cloth. Tom-Tom drum 6¼" diam. with cloth. "T" holds triangle and cymbal. Wooden bar holds 2 bells. Legs to hold bass drum and 2-8" sticks. Easily assembled with step by step instruction sheet. Each in corrugated case.
Packed ½ dozen minimum.
60 lbs. per dozen.

6

1953 catalogue showing "Hot Shot" drum sets.

DRUMS OF QUALITY

NO. 41 DRUM
Both heads metal. Lithoed in 4 colors. 6¼" diam. 2-8" sticks. 3 dox. ctn. 23 lbs.

NO. 42 DRUM
Heavy parchment heads. 18" neck sling. Lithoed in 4 colors. 2-12" sticks. 11" diam. 1 dox. ctn. 15 lbs.

NO. 44 DRUM
Bright litho as above but with all metal heads. 2 sticks and 18" neck sling. 3 shiny springs. 10" diam., 4½" deep.
1 dox. ctn. 15 lbs.

NO. 47 FIELD DRUM
Top head printed design. Bright metal lithoed 4 colors. 4 springs. Pair 12" sticks. 18" neck sling. 9" diam. 10" high.
½ dox. ctn. 11 lbs.

NO. 141 TOM-TOM
Bright "Indian" litho. Good year 'round. 3 springs. Beater w/knob 7½" long. 6¼" diam. 6½" high.
1 dox. ctn. 10½ lbs.

NO. 45 TRAP DRUM SET
Large drum 20½" diam. Body litho in bright red star design—rims blue. "Teenage Band" design in 3 colors. Drummer side a cloth head. Side head Snare drum 9" diam. Star design body—blue rims. 2-10". Tom-Tom lithoed in star design and has cloth head and is 6¼" diam. "T" bracket holds 7" cymbal, large triangle, cow bell and 3 aluminum bells—2", 2½" and 2½" size. Side cymbal played by foot beater which is made of steel and very sturdy. Foot beater and 2 leg supports securely snap on drum rim.
Each in case weighing 9 lbs.

12

1956 catalogue showing "Hep Cats" drum sets.

Small 1950s drum. $35-$50.

PROVED FAST SELLERS --- POPULARLY PRICED

NO. 126 DIME BANK

Shiny "Dime Bank" has side markings and openings to $5.00. When full, turn screw in sturdy top and dimes come out bottom. Quality item.
6 doz. ctn. 9 lbs. (1 doz. pkg.)

NO. 127 DIME BANK

Same as above but each item mounted on printed display card.
6 doz. ctn. 9 lbs.

MAIL BOX BANK

NO. 185 MAIL BOX BANK

NO. 78 GLOBE BANK

New 4" diam. size. Has lock and key. All bright 4 color litho.
4 doz. ctn. 14 lbs.

BRIGHTLY COLORED MAIL BOX BANK

NO. 185 MAIL BOX BANK

- WITH PADLOCK and KEYS
- STURDY METAL
- POPULAR SALES BUILDER

Really novel bank, brightly lithoed in red, white, blue and gold. Slot takes 50 cent piece. Front door held closed by real padlock with 2 keys. Kids love to lock up "treasurers". 5¼" high, 4" wide, 2¼" deep. Each in sleeve.
2 doz. ctn. 12 lbs.

YEAR 'ROUND SELLERS

NO. 121 $100,000 MONEY BOX

Sturdy metal brightly lithoed has recessed wire handle and spring catch to lock with 2 key sturdy padlock included. Cello pack of 45 play bills to $25,000 size. Cello front envelope of 18 aluminum coins in 5¢, 10¢, 25¢ etc. Lots of play-kids love to lock up secrets. Box is 8 1/8 x 3¾ x 1½". Each in printed sleeve.
1 doz. ctn. 9 lbs.

NO. 92 TOOL BOX W/VINYL TOOLS

Sturdy metal box, brightly lithoed is 8 1/8 x 3 3/4 x 1 1/2". Has recessed wire handle in top and spring catch to lock. Tools of soft, rubber-like Vinyl plastic. Saw 7¼" long, Pliers, Screwdriver, Combination Wrench, Hammer and Hatchet have wood handles. Each in sleeve.
1 doz. ctn. 11 lbs.

7

1956 catalogue page showing mail box bank, money box, and tool box.

Money box with all the play money. No. 121. 1956. *Courtesy of Bill Kerr.* $35-$60.

Two late 1950s tops. $35-$50.

1959 push toy. *Courtesy of Ohio Art Company.* $50-$85.

Two Bunny toys. The one in the foreground is from the 1930s and the other is the 1950s. Notice the difference in the quality and feeling of the lithography.

Holiday Bucket with candycane stripes. 1950s. Medium size. *Courtesy of Ohio Art Company.* $25-$40.

"ETCH A SKETCH"

The 1960s witnessed a dramatic achievement for Ohio Art—the world renowned "Etch A Sketch." In the late 1950s, a Parisian garage mechanic, Arthur Granjean, came up with the idea of a "magic screen." He took the idea to the 1959 Nuremberg toy fair in Germany where it was seen by a representative of Ohio Art. Later that same year, an employee of Ohio Art arranged for an Ohio Art executive to meet Arthur Granjean. The inventor came to Bryan and Howard Winzeler decided to take the risk and invest $25,000 in the licensing of "L'Ecran Magique." The screen appeared on toy shelves on July 12, 1960. It was renamed "Etch A Sketch."

Today, "Etch A Sketch" is the toy most associated with the Ohio Art name. Marketing studies have shown that approximately eight out of ten people are familiar with the "Etch A Sketch," while only three out of one hundred people are familiar with the Ohio Art Company.

"Etch A Sketch" was the first Ohio Art product advertised on television. It was promoted by Arthur Godfrey, Barbara Walters, Jean Shallit, and Woody Woodpecker, to name just a few.

Some of the famous celebrities who promoted the "Etch A Sketch."

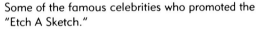

Some of the famous celebrities who promoted the "Etch A Sketch."

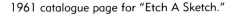

1961 catalogue page for "Etch A Sketch."

"Etch A Sketch" has received quite a bit of notoriety over the years. It was licensed internationally and was the subject of a case study at the University of Virginia MBA course. "Etch A Sketch" was the subject of a law suit initiated by Ralph Nader in the 1960s. He cited the toy as dangerous to small children. The story has it that the judge trying the case accidentally knocked the toy to the floor; it didn't break. Upon discovering its durability, he dismissed the case against Ohio Art.

"Etch A Sketch" is still produced in Bryan, Ohio, today. Nothing has changed since the 1960 model except for the size of the knobs and the production of several novelty items.

1967 catalogue page for "Etch A Sketch."

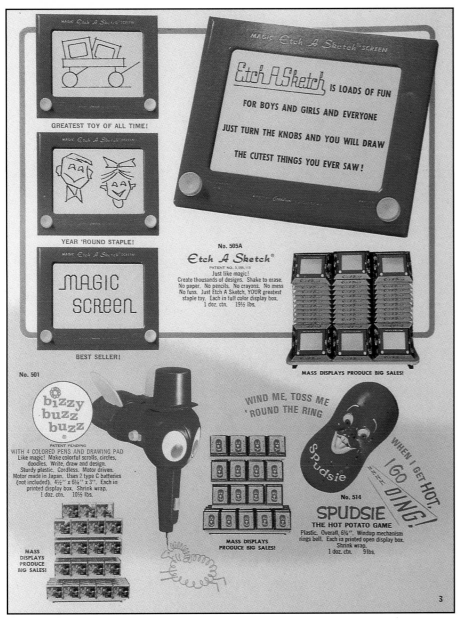

1968 catalogue page for "Etch A Sketch."

1969 catalogue advertisements for "Etch A Sketch." The 1969 catalogue shows a bird's eye view of the Ohio Art factory.

"Etch A Sketch" with a "Toy Story" theme. This is a much later "Etch A Sketch," produced just after the Disney movie was released.

1960s catalogue page showing several Technofix mechanical toys.

One could imagine that in today's world of electronic games and instant gratification, it would be hard to sustain an interest in a hand held sketching toy. However, the recent revival of interest in baby boomer toys and the block buster Disney movie, "Toy Story" contributed to the resurgence of interest in the "Etch A Sketch." In fact, the image of the "Etch A Sketch" was used in the promotion for all of the Toy Story advertisements. "Etch A Sketches" have been reduced to travel and keychain sizes.

OTHER TOYS

One of Ohio Art's most complex series of wind-up toys was a result of leasing dies from a German manufacturer called Technofix. Ohio Art bought the tooling from Technofix and produced no. 601, "Switch and Dump" from 1964-66; no. 611, "Traffic Control" from 1964 from 1964-65; no. 612, "Hi-Way Road Set" from 1964 through 1966; no 614, "Alpine Express" from 1965-1967; and no. 634, "Coney Island" from 1965 through 1967. All the small cars that were used on these products were imported from Western Germany. They definitely have a different look than the majority of Ohio Art toys.

Technofix "Coney Island" mechanical toy. *Courtesy of Ohio Art Company.* $100-$125.

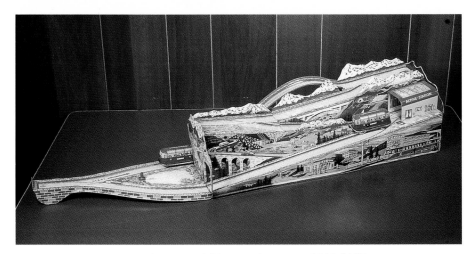

Technofix mechanical toy. *Courtesy of Ohio Art Company.* $100-$125.

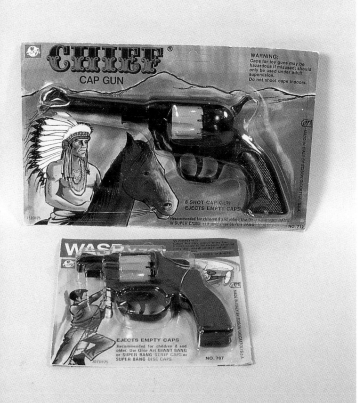

"Chief" gun and "Wasp" gun. $15-$25.

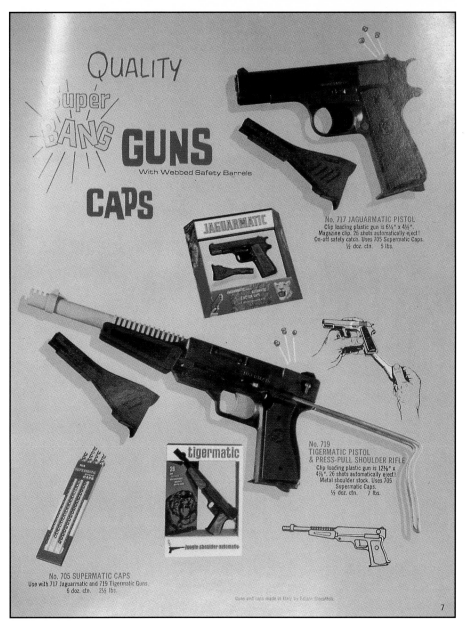

1967 catalogue advertising toy guns.

In 1965, Ohio Art started importing cap guns and caps from an Italian company, Edison Giacatolli. In 1980, Ohio Art decided to copy one of the Giocattoli guns and manufactured it themselves. With Bryan Die Cast Co. doing all the diecast parts, and Stryal, Ohio Art's injection molding plant, doing all the plastic parts, the new gun, called "The Wolf," was produced. It is the only gun marked "Made in Bryan, Ohio" on the blister card. In 1985, the guns were discontinued.

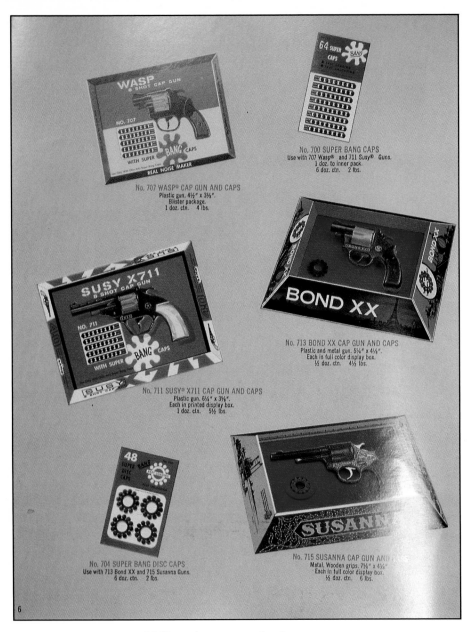

1967 catalogue page showing guns.

1962 catalogue pages showing various drums including two trap drum sets.

In 1968 the Ohio Art Company became the majority stockholder of the Emenee Corporation, a company known for production of children's musical instruments. Ohio Art started manufacturing instruments including keyboards, guitars, accordions, and saxophones. Through the years, Ohio Art licensed with the "Archies," "Smurfs," and "Care Bears." The company also manufactured "The Family Affair" organ based on the television show of the same name. In addition, the company continued to produce trap drum sets of various sizes.

Ohio Art Drums *up Big Sales for you!*

BASIN BEAT DRUM SET

No. 352 BASIN BEAT DRUM SET
- CLOSE WOUND DOUBLE WOUND SNARES.
- AMPLIFYING TONE BLOCK AND BELL.
- HEAVY DUTY FOOT PEDAL.
- EACH IN FULL COLOR BOX.
- BASS DRUM, 21½" diameter, 15½" deep.
- DRUM RINGS, 2".
- SNARE DRUM, 13" diameter, 5½" deep.
- TOM TOM, 11" diameter, 10½" deep.
- HIS-S-S-SING CYMBAL, 14".
- Drum playing instruction booklet.
- 1 Each 21 lbs.

Features, Features, FEATURES!
... and the quality it takes to
make them last!

- All playing heads are heavy duty frosted MYLAR®.
- Heavy gauge steel body.
- Rugged metal and high impact plastic pedal with heavy duty spring.
- Adjustable hi-tension drum clips.
- Nickel plated hardware.
- Pedal, 3" high, 10¾" long.
- Hardwood pro drumsticks, 15".

HEAVY DUTY FOOT PEDAL

TONE BLOCK AND BELL

NEW!

No. 336 TODDLER DRUM
- Starter drum.
- Lithoed metal.
- 2 wooden sticks.
- Each in printed self display sleeve.
- 6¼" diameter.
- 3¾" high.
- 1 doz. ctn. Approx. wt. 5½ lbs.

1969 catalogue page showing the "Basin Beat" drum set.

In 1968 only, Ohio Art bought a snap-apart plastic figure from "Timpo" in Great Britain. The figures— knights, cowboys, Indians, and blue and gray soldiers —were packaged at Ohio Art and sold under the Ohio Art name.

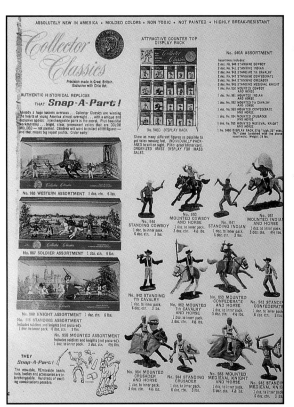

1968 catalogue showing toy soldiers.

Of course, there was the continued production of standard toys such as sand pails and tea sets, wind-ups, tops, and drums.

1963 catalogue showing targets.

Early 1960s target game. *Courtesy of Jim Gilcher.* $25-$40.

Late 1960s target game. *Courtesy of Jim Gilcher.* $25-$50.

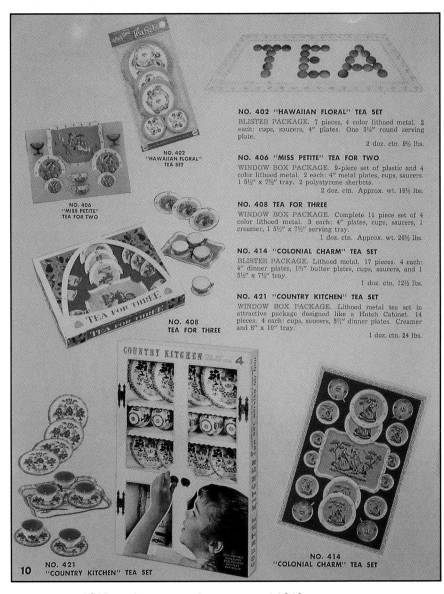

NO. 402 "HAWAIIAN FLORAL" TEA SET

BLISTER PACKAGE. 7 pieces, 4 color lithoed metal. 2 each: cups, saucers, 4" plates. One 5¼" round serving plate.

2 doz. ctn. 9½ lbs.

NO. 406 "MISS PETITE" TEA FOR TWO

WINDOW BOX PACKAGE. 9-piece set of plastic and 4 color lithoed metal. 2 each: 4" metal plates, cups, saucers. 1 5½" x 7½" tray. 2 polystyrene sherbets.

2 doz. ctn. Approx. wt. 19½ lbs.

NO. 408 TEA FOR THREE

WINDOW BOX PACKAGE. Complete 11 piece set of 4 color lithoed metal. 3 each: 4" plates, cups, saucers, 1 creamer, 1 5½" x 7½" serving tray.

1 doz. ctn. Approx. wt. 26½ lbs.

NO. 414 "COLONIAL CHARM" TEA SET

BLISTER PACKAGE. Lithoed metal. 17 pieces. 4 each: 4" dinner plates, 1¾" butter plates, cups, saucers, and 1 5¼" x 7½" tray.

1 doz. ctn. 12½ lbs.

NO. 421 "COUNTRY KITCHEN" TEA SET

WINDOW BOX PACKAGE. Lithoed metal tea set in attractive package designed like a Hutch Cabinet. 14 pieces. 4 each: cups, saucers, 5¼" dinner plates. Creamer and 8" x 10" tray.

1 doz. ctn. 24 lbs.

1965 catalogue page showing typical 1960s tea sets.

Another 1960s "Little Red Riding Hood" tea set. Note pattern on the bottom of the cups.

"Little Red Riding Hood" tea set. 1960s. *Courtesy of Jim Gilcher.* Eleven-piece set: $50-$85.

"Pink Poodle" tea set. Early 1960s. Eleven-piece set: $55-$85.

"Three Little Pigs" tea set. Late 1960s. Eleven-piece set: $45-$65.

"Swiss Miss" tea set. 1967. Thirteen-piece set: $50-$75.

Plastic tea set. 1960s. *Courtesy of Jim Gilcher.* Twenty-six piece set on card: $40.

BRILLIANT
SWISS MUSIC TOP

● GENUINE SWISS MUSIC UNIT

● PLAYS COMPLETE TUNE

Really a quality item. Bell shaped and beautifully decorated in gold and red, blue and green iridescent colors. Has "old world" look. Easy running. Swiss unit plays complete tune only when top is spun with spiral. Gold wooden Knob. 9½" high, 8½" diam. Beautiful display box. ½ doz. ctn. 14 lbs.

NO. 123 SWISS MUSIC TOP

AUTOMATIC
CHORAL TONE TOP

NO. 37 AUTOMATIC CHORAL TOP

We are really proud to be the first to make in this country a real "Automatic Choral" Musical Top. Item has a 12-tone Choral reed and changes tones—making beautiful chords—while spinning. Four-color litho. Rubber suction base attached. 10¾" high, 8½" diameter. Each in printed box. ½ doz. ctn. 11 lbs.

Base of top now rubber suction type

CONTINUOUS
COLOR
CHANGER

The 3 color disc make a rainbow of colors as top is spinning. Musical reeds play at same time. Colors keep changing as long as top spins on rubber base. Clear plastic top shell-very sturdy & well made. Plunger type spiral. 10" high, 7½" diam. Ea. in box. ½ doz. ctn. 11 lbs.

NO. 124 MUSICAL, COLOR TOP

NO. 33 TOP

4 color litho. Spiral wind. 5¼" diam. 5¾" high. 3 doz. ctn. 13 lbs.

NO. 33 TOP

NO. 35 MUSICAL TOP

Plunger type. 4 color litho. 7" diam. 8" high. 3 tone reed. Ea. 2 doz. ctn. 18 lbs.

NO. 36 MUSICAL TOP

Plunger type. 4 color litho. 4 tone reed. 9½" diam. 9" high. Each in box. 1 doz. ctn. 15 lbs.

NO. 122 MUSICAL, COLOR CHANGING TOP

Bright litho colors, clear plastic top shows "Rainbow" changing colors when top is spun, also playing musical reed. 7½" high, 7" diam. Ea. in beautiful box. 1 doz. ctn. 13 lbs.

14

"Here We Go Round the Mulberry Bush" top. Late 1960s. $25-$45.

Musical top. Produced throughout the 1960s. $25-45.

Tiger pail. Late 1960s. Medium. $30-$45.

Flowered pail with little boy and girl. Late 1950s-early 1960s. Large. $45-$60.

Green iridescent fish pail.
1960s. Medium. $35-$50.

Iridescent Parrots sand pail. Late 1960s.
Medium. $45-$60.

Iridescent magenta fish pail. Medium.
$40-$55.

Iridescent Eagle sand pail. 1960s. *Courtesy of Claudette Job.* Medium. $50-$75.

Iridescent Toucans sand pail. Early 1960s. Medium. $45-$60.

One of many globes produced by Ohio Art. This one dates from 1964. It has the alphabet, numbers, and animals. *Courtesy of John Walters.*

"Astro Ray" Dart Gun. 1968-70. *Courtesy of Ohio Art Company.* $100-$150.

1968 catalogue showing action figures.

Marble Game. 1965. *Courtesy of Jim Gilcher.* $25-$40.

No. 562 ASTRORAY GUN
Flashlight gun lights up target! This new eight piece dart set is for tomorrow's space aces. Red and white plastic Astroray gun features adjustable "Ray" for accuracy in hitting target. Safe—operates from two size "c" flashlight batteries (not included). Bulb included. 12¾" diameter metal target in four color litho. Six assorted colored safe plastic darts with rubber suction cup tips. Gun is 9¾" x 7½" x 2½". Each in open display box. Packed 1 doz. to ctn. Approx. wt. 22⅛ lbs.

No. 364 FIDO'S MUSICAL DOG HOUSE WITH GENUINE SWISS MUSICAL UNIT
Genuine Swiss musical unit plays familiar "Where, Oh Where Has My Little Dog Gone" until Fido "jumps" from dog house door. Four color lithoed metal dog house measures 7¼" x 7½" x 7¾". Wire handle for carrying. Three colored plastic dog. Boxed 1 each. Packed 1 doz. to ctn. Approx. wt. 23 lbs.

No. 503 SKETCH A GRAPH
With Pantograph
An entirely new concept of the pantograph in a practical toy. Enlarge or reduce almost any drawing . . . in color . . . in perfect proportion. Four arms, 3 different scales for different sizes. Three complete pens, one each: red, green, blue. Suction cups keep original copy flat, eliminate slipping. Set includes red plastic Sketch A Graph, 3 pens, drawing paper and sketch book. Each in full color box. Size box 17" x 14¼" x 1¾". Packed 1 doz. to ctn. Approx. wt. 28 lbs.

ART

3

1962 catalogue showing various toys.

1967 catalogue showing garden toys.

Boomerang. 1963-64. *Courtesy of Jim Gilcher.*
$5-$15.

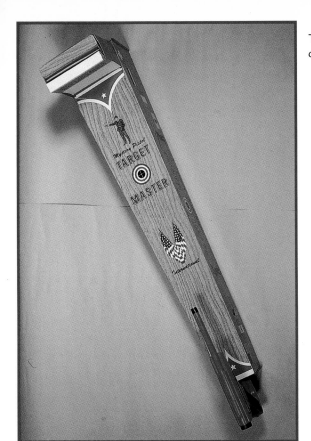

Target game. 1962-63. *Courtesy of Jim Gilcher.*

Barbecue trays. Set of five. 1964-65. *Courtesy of Jim Gilcher.* $125-$250.

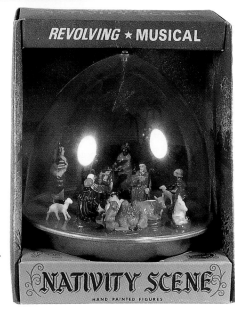

Nativity Scene. $35-$60.

CHAPTER VII:
THE SEVENTIES AND EIGHTIES

Throughout the 1970s and 1980s Ohio Art continued to produce toys. The lithography became progressively more cartoon-like. Gone was the subtle complexity of the 1920s and 1930s and the cute, animated primary colors of the 1940s and 1950s. Instead, larger plain blocks of intense "hot" colors were laid down reflecting the often "charmless" animation of the 1970s and 1980s.

The tea sets and housekeeping toys of the 1970s and 1980s came under trademarks such as "Sunnie Miss," and "Pfaltzgraff." For boys, the company produced baseballs, footballs, basketball hoops, and yo-yos.

1981 catalogue page showing assorted "Sunnie Miss" tea sets and a toaster.

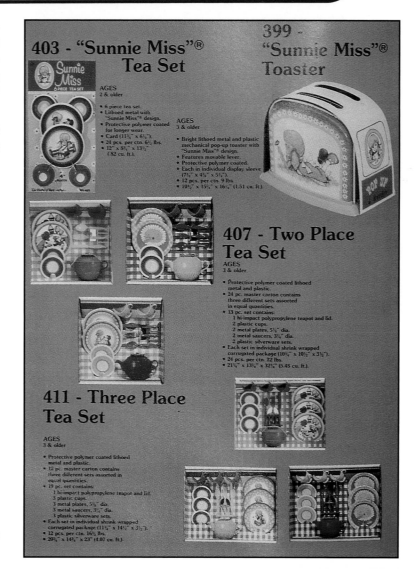

403 - "Sunnie Miss"® Tea Set

399 - "Sunnie Miss"® Toaster

AGES 2 & older

- 6 piece tea set.
- Lithoed metal with "Sunnie Miss"® design.
- Protective polymer coated for longer wear.
- Card (11¾" x 6½").
- 24 pcs. per ctn. 6½ lbs.
- 12" x 5½" x 13½" (.82 cu. ft.).

AGES 3 & older

- Bright lithoed metal and plastic mechanical pop-up toaster with "Sunnie Miss"® design.
- Features movable lever.
- Protective polymer coated.
- Each in individual display sleeve (7½" x 4½" x 5½").
- 12 pcs. per ctn. 9 lbs
- 10½" x 15¼" x 16¼" (1.51 cu. ft.)

407 - Two Place Tea Set

AGES 3 & older

- Protective polymer coated lithoed metal and plastic.
- 24 pc. master carton contains three different sets assorted in equal quantities.
- 13 pc. set contains:
  1 hi-impact polypropylene teapot and lid.
  2 plastic cups.
  2 metal plates, 5½" dia.
  2 metal saucers, 3¾" dia.
  2 plastic silverware sets.
- Each set in individual shrink wrapped corrugated package (10¼" x 10½" x 3½").
- 24 pcs. per ctn. 22 lbs.
- 21¾" x 13¾" x 32¾" (5.45 cu. ft.)

411 - Three Place Tea Set

AGES 3 & older

- Protective polymer coated lithoed metal and plastic.
- 12 pc. master carton contains three different sets assorted in equal quantities.
- 19 pc. set contains:
  1 hi-impact polypropylene teapot and lid.
  3 plastic cups.
  3 metal plates, 5½" dia.
  3 metal saucers, 3¾" dia.
  3 plastic silverware sets.
- Each set in individual shrink wrapped corrugated package (11¾" x 14½" x 3½").
- 12 pcs. per ctn. 16½ lbs.
- 20½" x 14¾" x 23" (4.07 cu. ft.).

1984 catalogue page showing "Pfaltzgraff" tea sets and "Charlie Brown" and "Snoopy" sets.

1979 catalogue page showing yo-yos.

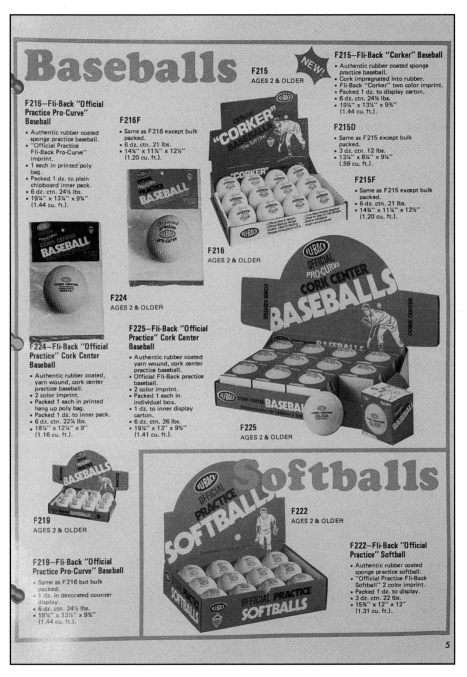

Baseballs

F215
AGES 2 & OLDER

NEW!

F215—Fli-Back "Corker" Baseball
- Authentic rubber coated sponge practice baseball.
- Cork impregnated into rubber.
- Fli-Back "Corker" two color imprint.
- Packed 1 dz. to display carton.
- 6 dz. ctn. 24½ lbs.
- 19¼" x 13¼" x 9¾" (1.44 cu. ft.).

F216—Fli-Back "Official Practice Pro-Curve" Baseball
- Authentic rubber coated sponge practice baseball.
- "Official Practice Fli-Back Pro-Curve" imprint.
- 1 each in printed poly bag.
- Packed 1 dz. to plain chipboard inner pack.
- 6 dz. ctn. 24½ lbs.
- 19¼" x 13¼" x 9¾" (1.44 cu. ft.).

F216F
- Same as F216 except bulk packed.
- 6 dz. ctn. 21 lbs.
- 14¾" x 11¼" x 12½" (1.20 cu. ft.).

F215D
- Same as F215 except bulk packed.
- 3 dz. ctn. 12 lbs.
- 13¼" x 8¼" x 9¼" (.59 cu. ft.).

F215F
- Same as F215 except bulk packed.
- 6 dz. ctn. 21 lbs.
- 14¾" x 11¼" x 12½" (1.20 cu. ft.).

F216
AGES 2 & OLDER

F224
AGES 2 & OLDER

F224—Fli-Back "Official Practice" Cork Center Baseball
- Authentic rubber coated, yarn wound, cork center practice baseball.
- 2 color imprint.
- Packed 1 each in printed hang up poly bag.
- Packed 1 dz. to inner pack.
- 6 dz. ctn. 22¼ lbs.
- 18¼" x 12¼" x 9" (1.16 cu. ft.).

F225—Fli-Back "Official Practice" Cork Center Baseball
- Authentic rubber coated yarn wound, cork center practice baseball.
- Official Fli-Back practice baseball.
- 2 color imprint.
- Packed 1 each in individual box.
- 1 dz. to inner display carton.
- 6 dz. ctn. 26 lbs.
- 19¼" x 13" x 9¾" (1.41 cu. ft.).

F225
AGES 2 & OLDER

Softballs

F219
AGES 2 & OLDER

F219—Fli-Back "Official Practice Pro-Curve" Baseball
- Same as F216 but bulk packed.
- 1 dz. in decorated counter display.
- 6 dz. ctn. 24½ lbs.
- 19¼" x 13¼" x 9¾" (1.44 cu. ft.).

F222
AGES 2 & OLDER

F222—Fli-Back "Official Practice" Softball
- Authentic rubber coated sponge practice softball.
- "Official Practice Fli-Back Softball" 2 color imprint.
- Packed 1 dz. to display.
- 3 dz. ctn. 22 lbs.
- 15¾" x 12" x 12" (1.31 cu. ft.).

5

1979 catalogue page showing baseballs.

646 - John Havlicek "Dribbler"™ Basketball AGES 4 & older
- Super bounce foam ball made from "Tuffoam"™.
- The look of a real basketball.
- Autographed by NBA Basketball Superstar John Havlicek.
- Shrink wrapped in decorated counter kraft corrugated display carton featuring John Havlicek's picture (14½" x 21½" x 14¼").
- 12 pcs. per ctn. 7 lbs.
- 15" x 14½" x 22" (2.76 cu. ft.).

650 - Lil' Sport® "Punt, Pass & Kick"™ Football AGES 4 & older
- Durable foam football made from "Tuffoam"™.
- 5½" dia. x 9½" long.
- Simulated laces and seams, Lil' Sport® logo.
- Shrink wrapped in decorated counter kraft corrugated display carton (16¼" x 18½" x 12").
- 12 pcs. per ctn. 7 lbs.
- 16¼" x 12½" x 19½" (2.36 cu. ft.).

644 - Lil' Sport® Playball
AGES 1 & older
- Soft foam 6⅜" dia. ball in bright yellow color.
- Great for indoor and outdoor fun.
- Each in individual chipboard display box (7" x 7" x 7").
- 12 pcs. per ctn. 8 lbs.
- 14" x 14¼" x 21½" (2.48 cu. ft.).

648 - Lil' Sport® Football
AGES 1 & older
- Foam ball in bright yellow color.
- Great for indoor and outdoor fun.
- 5" dia. x 8¾" long.
- Simulated laces and seams, Lil' Sport® logo.
- Each in individual chipboard display box (5¼" x 5¾" x 8").
- 12 pcs. per ctn. 6 lbs.
- 16¼" x 11" x 18" (1.86 cu. ft.).

10

1981 catalogue page showing endorsed basketballs and "Lil' Sport" balls.

As Ohio Art diversified and acquired other companies, the variety of toys bearing the Ohio Art name changed as well. An electric and battery operated sewing machine, the sturdy plastic "Tom Thumb" typewriter from England, and the "Tom Thumb" cash register all appeared in Ohio Art catalogues.

Ohio Art continued to market musical instruments including full-standing organs, trap drums sets, and electric guitars. In 1981 a line of "Pedigree" dolls appeared in catalogues.

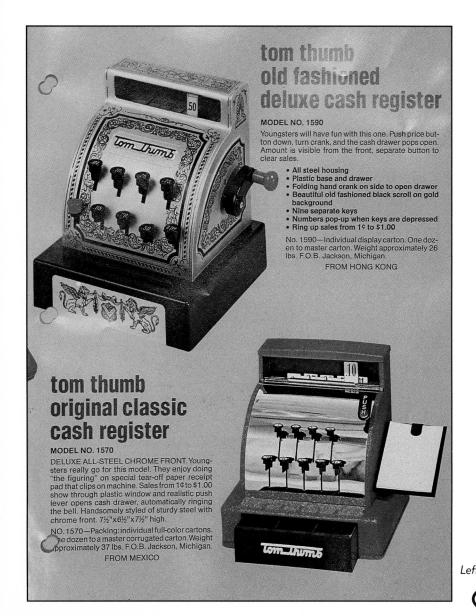

tom thumb old fashioned deluxe cash register

MODEL NO. 1590

Youngsters will have fun with this one. Push price button down, turn crank, and the cash drawer pops open. Amount is visible from the front, separate button to clear sales.

- All steel housing
- Plastic base and drawer
- Folding hand crank on side to open drawer
- Beautiful old fashioned black scroll on gold background
- Nine separate keys
- Numbers pop-up when keys are depressed
- Ring up sales from 1¢ to $1.00

No. 1590—Individual display carton. One dozen to master carton. Weight approximately 26 lbs. F.O.B. Jackson, Michigan.

FROM HONG KONG

tom thumb original classic cash register

MODEL NO. 1570

DELUXE ALL-STEEL CHROME FRONT. Youngsters really go for this model. They enjoy doing "the figuring" on special tear-off paper receipt pad that clips on machine. Sales from 1¢ to $1.00 show through plastic window and realistic push lever opens cash drawer, automatically ringing the bell. Handsomely styled of sturdy steel with chrome front. 7½"x6½"x7½" high.

NO. 1570—Packing: individual full-color cartons. One dozen to a master corrugated carton. Weight approximately 37 lbs. F.O.B. Jackson, Michigan.

FROM MEXICO

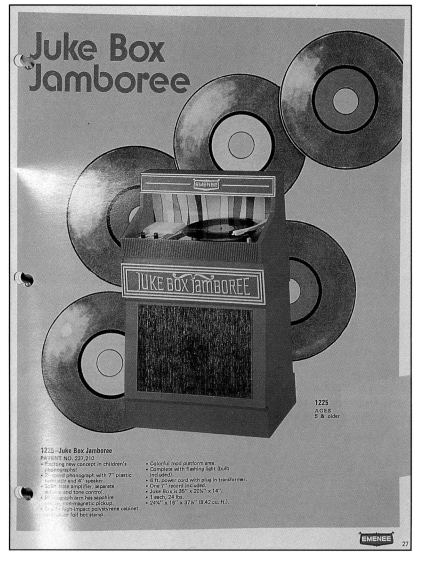

Juke Box Jamboree

1225—Juke Box Jamboree
PATENT NO. 237,210

- Exciting new concept in children's phonographs!
- 2-speed phonograph with 7" plastic turntable and 4" speaker.
- Solid state amplifier; separate volume and tone control.
- Phonograph arm has sapphire stylus, non-magnetic pickup.
- Sturdy high-impact polystyrene cabinet with silver foil hot stamp.
- Colorful mod platform area.
- Complete with flashing light (bulb included).
- 6 ft. power cord with plug in transformer.
- One 7" record included.
- Juke Box is 35" x 20½" x 14".
- 1 each, 24 lbs.
- 24¼" x 16" x 37½" (8.42 cu. ft.).

1225
AGES
5 & older

EMENEE
27

Above: 1978 catalogue page showing "Juke Box Jamboree."

Left: 1979 catalogue page showing "Tom Thumb" cash registers.

Ohio Art continued to turn out tin-litho sand pails, shovels, metal trays, tops, drums, holiday items, garden toys, housekeeping toys, Chinese Checkers sets, target games, and globes.

Variety of Ohio Art Chinese Checkers boards. *Courtesy of Peg and Rich Jansen.*

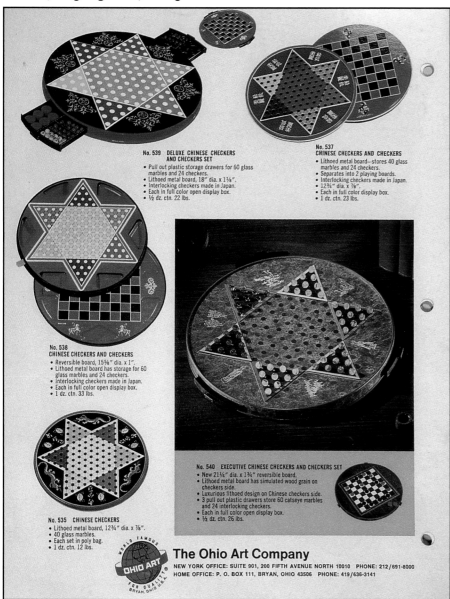

No. 539 DELUXE CHINESE CHECKERS AND CHECKERS SET
- Pull out plastic storage drawers for 60 glass marbles and 24 checkers.
- Lithoed metal board, 18" dia. x 1⅛".
- Interlocking checkers made in Japan.
- Each in full color open display box.
- ½ dz. ctn. 22 lbs.

No. 537 CHINESE CHECKERS AND CHECKERS
- Lithoed metal board—stores 40 glass marbles and 24 checkers.
- Separates into 2 playing boards.
- Interlocking checkers made in Japan.
- 12¾" dia. x ⅞".
- Each in full color display box.
- 1 dz. ctn. 23 lbs.

No. 538 CHINESE CHECKERS AND CHECKERS
- Reversible board, 15⅝" dia. x 1".
- Lithoed metal board has storage for 60 glass marbles and 24 checkers.
- Interlocking checkers made in Japan.
- Each in full color open display box.
- 1 dz. ctn. 33 lbs.

No. 540 EXECUTIVE CHINESE CHECKERS AND CHECKERS SET
- New 21¼" dia. x 1¾" reversible board.
- Lithoed metal board has simulated wood grain on checkers side.
- Luxurious lithoed design on Chinese checkers side.
- 3 pull out plastic drawers store 60 catseye marbles and 24 interlocking checkers.
- Each in full color open display box.
- ½ dz. ctn. 26 lbs.

No. 535 CHINESE CHECKERS
- Lithoed metal board, 12¾" dia. x ⅞".
- 40 glass marbles.
- Each set in poly bag.
- 1 dz. ctn. 12 lbs.

The Ohio Art Company
NEW YORK OFFICE: SUITE 901, 200 FIFTH AVENUE NORTH 10010   PHONE: 212/691-8000
HOME OFFICE: P. O. BOX 111, BRYAN, OHIO 43506   PHONE: 419/636-3141

1970s catalogue showing a variety of Chinese Checkers sets produced throughout the 1960s and 1970s.

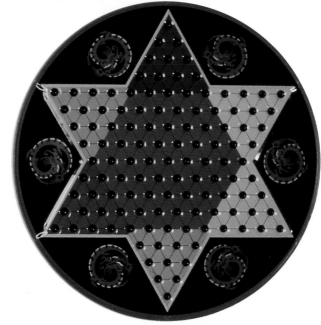

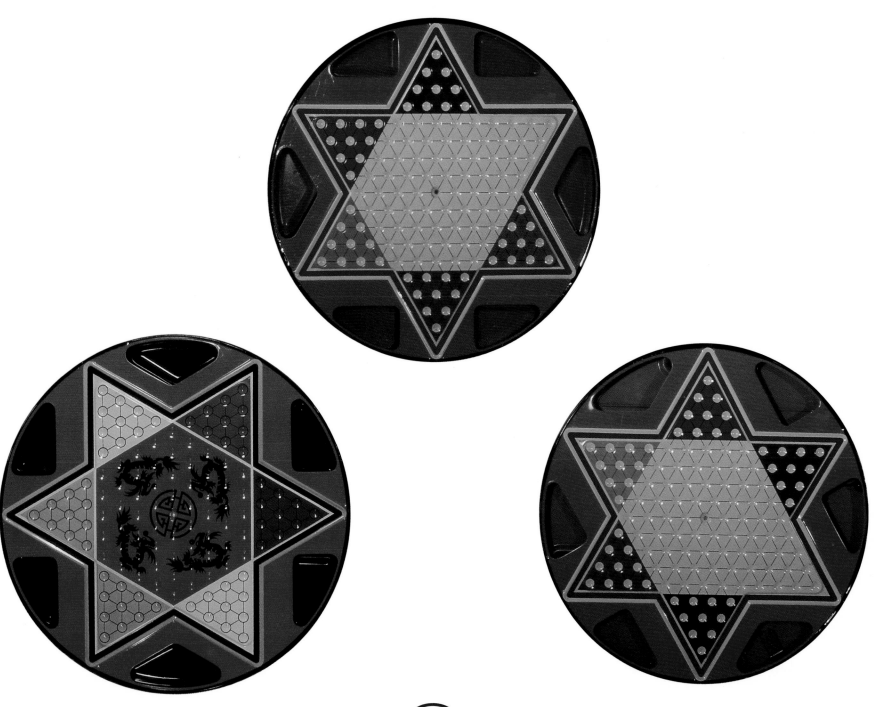

"Americana" tea set. 1974. *Courtesy of Ohio Art Company.* Eleven-piece set on card: $35.

"Storytime" tea set. 1973-74. *Courtesy of Ohio Art Company.* Thirteen-piece set on card: $50.

"Alice in Wonderland" tea set. 1982. *Courtesy of Jim Gilcher.* Fifteen-piece set on card: $35.

"Sunnie Miss" tea set. 1976. *Courtesy of Jim Gilcher.* Six-piece set on card: $20.

A variety of lunch boxes produced by Ohio Art in the 1970s. The flag was produced in 1973; the watermelon in 1974; and the ghost in 1977.

1971 catalogue cover showing tin-litho sand pails still in production.

"Little Red Riding Hood" sand pail. 1972. 8" tall. *Courtesy of Doug and Pat Wengel.* $30-$45.

"Anchor" sand pail. 1972. *Courtesy of Doug and Pat Wengel.* $30-$45.

"How to Fish" sand pail. 1972-78. *Courtesy of Doug and Pat Wengel.* $30-$45.

1972 catalogue page showing iridescent sand pails.

Iridescent turtle pail. 1970s. *Courtesy of Jim Gilcher*. Medium. $30-$45.

Two promotional pails. The one on the left is from the 1950s and the Hershey's one on the left dates from the 1970s. *Courtesy of Jim Gilcher.*

Four globes. Brown base: 1938; Light blue bases: 1941; dark blue base: 1974-75. *Courtesy of Ohio Art Company.*

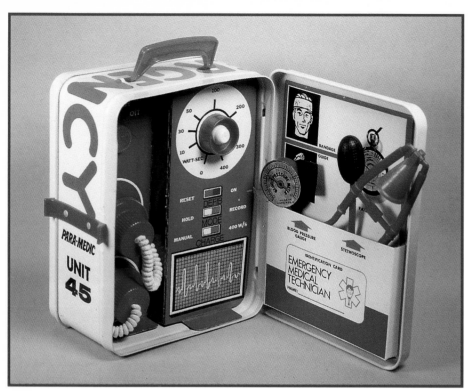

"Emergency Medical Kit." 1976. The other side of this kit opens up and has emergency telephone inside. This item was very popular. $40-$60. *Courtesy of Jim Gilcher.*

1979 catalogue showing paddle balls and Frisbees.

A 1960 Christmas tray and a 1971 paddle bat.

1983 catalogue with popular endorsements.

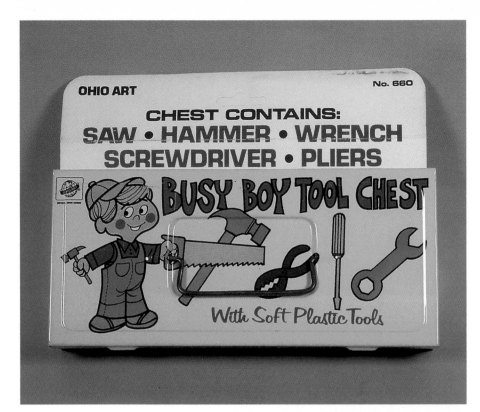

OHIO ART 1984

1984 catalogue showing kids' basketball hoops which are still in production today; player endorsements enhance their appeal.

OHIO ART No. 660

CHEST CONTAINS:
SAW • HAMMER • WRENCH
SCREWDRIVER • PLIERS

BUSY BOY TOOL CHEST

With Soft Plastic Tools

Tool kit. 1980. *Courtesy of Jim Gilcher.* $12-$16.

Two boats. The sail boat is from the 1940s and is contrasted with the 1980s speedboat. *Courtesy of Jim Gilcher.*

COAST GUARD

98-F175

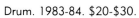
Drum. 1983-84. $20-$30.

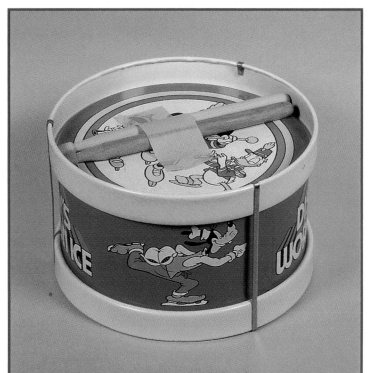

Sprinkling
can. 1986.
$20-$30.

"Bozo the Clown" top.
1980. $35-$60. *Courtesy of Jim Gilcher.*

Many Ohio Art toys enjoyed success in the marketplace, but as is true of any successful company, there are always a few bloopers.

"Western Woolie Walkers." 1982. These toys, imported from Hong Kong, were not a big hit and were quickly discontinued. *Courtesy of Jim Gilcher.*

"Sport Pets." 1983. There were nine different pets, made in Hong Kong. These were discontinued after one year. *Courtesy of Jim Gilcher.*

At this point in time, any toys produced after 1980 are not considered collectible. However, watch out for 2020 and hang onto those toys!

Opposite page: Promotional item for Mickey Mouse "Etch A Sketch." Circa 1989. Due to the expense of Disney licensing, the Mickey Mouse "Etch A Sketch" sold for $20 compared to the generic "Etch A Sketch" which cost only $10. It did not sell very well here in the United States but did sell fairly well in France. Only twelve of these huge plastic Mickey Mouse "Etch A Sketch" promotional items were produced; six went overseas. The battery-operated eyes move back and forth.

"Snoopy" board game. This prototype went to the New York Toy Fair in 1984. Due to a lack of interest, it never went into production.

FOR MORE INFORMATION

For those of you who can't get enough, here is a list of some clubs and individuals you can contact for more information:

Ohio Art Collectors Club
Sharone (Sherry) Lazane
Sharon Lazane
Ohio Art Beat
Quarterly Newsletters
Membership List
Buy/Sale and Trade Ads for members
18203 Kristi Road West
Liberty, Missouri, 64068
816-781-5452
e-mail: slazane@aol.com

The Ohio Art Company
The Etch A Sketch Club
1 Toy Street
P.O. Box 111
Bryan, OH
43506

Toy Dish Collectors Club
Shelley Smith
P.O. Box 159
Bethlehem, CT